Amphibians & Reptiles

of

Minnesota, Wisconsin & Michigan

Allen Blake Sheldon

Kollath+Stensaas
PUBLISHING

Kollath+Stensaas Publishing
394 Lake Avenue South, Suite 406
Duluth, MN 55802
Office: 218.727.1731
Orders: 1.800.678.7006 (Adventure Keen Distributors)
info@kollathstensaas.com
www.kollathstensaas.com

AMPHIBIANS & REPTILES
of MINNESOTA, WISCONSIN & MICHIGAN

Printed in U.S.A by J.S. Print Group
10 9 8 7 6 5 4 3 2 1 First Edition

Editorial Director: Mark Sparky Stensaas
Graphic Design: Kollath+Stensaas Publishing
Illustrations by Rick Kollath [www.kollathdesign.com]

ISBN-13: 978-1-936571-14-7

Table of Contents

To Beth & Lynn,
who stopped highway traffic
to save a Snapping Turtle

Acknowledgements

First, I thank Sparky Stensaas and Rick Kollath for having the confidence to select me to write this book, especially considering the impressive quality of their previously published nature field guides.

I'm grateful to my daughters, Beth and Lynn, and their mother, Kathy, for enduring occasional inconveniences, but still encouraging my interest in photographing amphibians and reptiles.

Many of the amphibians and reptiles pictured in this book were provided by my friends Dan Nedrelo and Bob Hay, or else found during our quests into the forests and wetlands of Wisconsin. They have inspired me to get out in the field more often, and we've waded through chilly marshes in the middle of the night, dragged canoes through shallow streams, and flipped innumerable rocks in search of elusive critters.

Matt Heeter accompanied me often to observe and learn more about rattlesnakes and Terry Odegaard hunted snakes with me when we were in eighth grade and we still look for them today.

Amphibians and reptiles have also been made available by the enthusiastic and knowledgeable members of the Minnesota Herpetological Society, who have brought critters to meetings and picnics or captured them during surveys.

My thanks to the sharp-eyed members of the La Crosse Area Camera Club, who helped improve my photography with 40 years of ruthless criticism.

Thanks again, and I hope everyone enjoys the book.

Allen Blake Sheldon
August 25, 2021

We at Kollath-Stensaas have been blessed with great authors; Allen continues that tradition. He met every deadline with quality work; work that sprang from his incredible knowledge of our creeping and crawling critters. Thanks Allen.

The Publishers; Mark Sparky Stensaas, Rick Kollath
September 2021

Why Should I be Interested in Creepy Crawly Critters?

Amphibians and reptiles are an integral part of Minnesota, Wisconsin and Michigan, adding to our enjoyment of nature. We might only get a glimpse as frogs leap from shore, turtles tumble off logs and splash into the water, or a snake zips across a trail. But unlike most birds and mammals, some amphibians and reptiles can be closely observed. Instead of fleeing, a harmless hog-nosed snake might coil and treat us to a lively threat display followed by its death act. A tiny bright green treefrog crouched on a green leaf, relying on its cryptic coloration to escape detection, usually remains motionless while we observe it. We also enjoy hearing frogs and toads, as their breeding choruses rival the cry of the loon for enhancing our night.

If you familiarize yourself with the snakes inhabiting our area, it will cause you less anxiety when you encounter one. Instead of trying to remember general rules about distinguishing a venomous snake from a harmless one, you can just say, "It's only a foxsnake."

Reptiles and amphibians are significant strands in the food web. We obviously benefit when an army of frogs is snapping up pesky insects, or when snakes help control populations of mice and voles. In a different role amphibians and reptiles, especially in the form of frogs, tadpoles, and turtle eggs, are important food sources of many species of wildlife.

Amphibians have been recognized as being valuable indicators of environmental health.

Amphibians, with their unshelled eggs and permeable skin, are more susceptible to pollution and other environmental deterioration than most other animals. Also, because most amphibians have life cycles that require living both in water and on land, they are vulnerable to problems with either habitat. There is a good chance that a new, unexpected environmental threat would impact amphibians first, giving us a warning signal before we are affected. Unfortunately, in recent years, unprecedented numbers of our little amphibian warning signals have been found grotesquely deformed or not found at all.

Many populations have decreased or disappeared for no apparent reason. Something is obviously wrong and will be discussed in a later chapter.

Milksnake

What Are Amphibians and Reptiles?

Amphibians and reptiles are ectothermic or cold-blooded, meaning they do not regulate body temperature internally as do the endothermic birds and mammals. Instead, amphibians and reptiles control their body temperature behaviorally by changing locations or even their postures while sunning. Reptiles may frequently move in and out of the sun to maintain an optimum body temperature. For a lizard that's been out in the sun for a while, it seems inappropriate to call it cold-blooded, so ectothermic is a better term. Amphibians spend less effort regulating body temperature, because their moist skin is better maintained in cool, moist places.

The cold climate of our three northern states forces amphibians and reptiles, except for a couple aquatic salamanders, into hibernation (sometimes called brumation for amphibians and reptiles) for roughly half the year. Various survival methods are used. Some frogs and turtles retreat to the bottom of the lake or river. Some snakes gather at hibernacula, such as dens on southwest-facing slopes. Other snakes use anthills or crayfish burrows. Many amphibians just find shelter individually. Most species retreat below the frost level, while a few have actually adapted enough to survive freezing temperatures.

Crawling, hopping, slithering, or other amphibian and reptile locomotion doesn't cover a lot of ground very fast, and the home territory of any individual is not large. Similarly, the geographic range of any species is fairly well defined and not rapidly expanded.

Most amphibians and reptiles periodically shed their outer layer of skin to accommodate growth, a process called ecdysis. Some shed the whole skin at once, while others shed it in pieces.

On the vertebrate evolutionary scale, amphibians and reptiles are above the fish, and below the birds and mammals. Amphibians and reptiles have obvious external differences: Amphibians have no scales or bony plates, toes are not clawed, and the skin is moist or slimy. Reptiles have scales or bony plates, claws on most toes, and normally dry skin.

Herpetology

The study of amphibians and reptiles is called herpetology, from the Greek word "herpeton," which means to creep or crawl. Herpetology buffs have coined and regularly use several colloquial offshoots: herpetofauna or herps are amphibians and reptiles, herping is searching for them, and herpetoculture emphasizes pets and captive breeding. Obviously it's easier to say "herps" than "amphibians and reptiles." It's certainly easier to type!

Classification of Herps

Carolus Linnaeus

To better understand the evolutionary and current relationships of our amphibians and reptiles, it helps to know how they are classified. Every organism, living or extinct, is placed into categories based on structural similarities or differences with other organisms. This process is called taxonomy. The taxonomic system used today was devised in 1735 by the Swedish scientist, Karl von Linne', better known by his Latinized name, Carolus Linnaeus. He also began a system of binomial nomenclature: assigning a unique pair of Latin names to every species of animal and plant. The pair of Latin names, called the scientific name, identifies the species to scientists, regardless of what language they speak.

Consider the Tiger Salamander. Along with every other animal on the planet, the Tiger Salamander is in the kingdom *Animalia.* Animals are divided into about 35 groups called *phyla.* The Tiger Salamander is in the phylum Chordata, which contains all the mammals, birds, reptiles, amphibians, and fish, all of which have a spinal cord. Chordata is divided into seven classes: *Mammalia, Aves, Reptilia, Amphibia,* and three classes of fish. The class Amphibia, which includes the Tiger Salamander, is divided into orders. Salamanders are in the order Caudata, which is divided into families. The Tiger Salamander is in the family Ambystomatidae, which is the group of salamanders called mole salamanders, because they burrow and usually live underground. Families are divided into genera, with the Tiger Salamander given the genus name *Ambystoma.* Genus names are always italicized and the first letter is always capitalized. Five other salamanders in our three states are closely related and have the same genus name. Different specific epithets, also italicized, finally distinguish them. The Tiger Salamander is *Ambystoma tigrinum,* which is its scientific name. Subspecies are given an additional subspecific epithet. In our area there is one subspecies of Tiger Salamander: Eastern Tiger Salamander *Ambystoma tigrinum tigrinum.*

Tiger Salamander

Meet the Amphibians

Amphibian is a combination of two Greek words: *amphi* meaning double and *bios* meaning life. As the name states, many amphibian species lead two lives: first as an aquatic form breathing with gills, and then as a terrestrial form breathing with lungs. But there are many exceptions to this generalization.

Amphibian External Features

Adult Frog Features

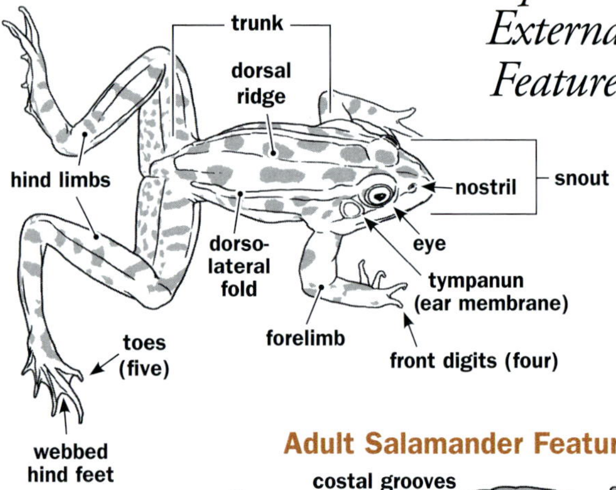

Adult Salamander Features

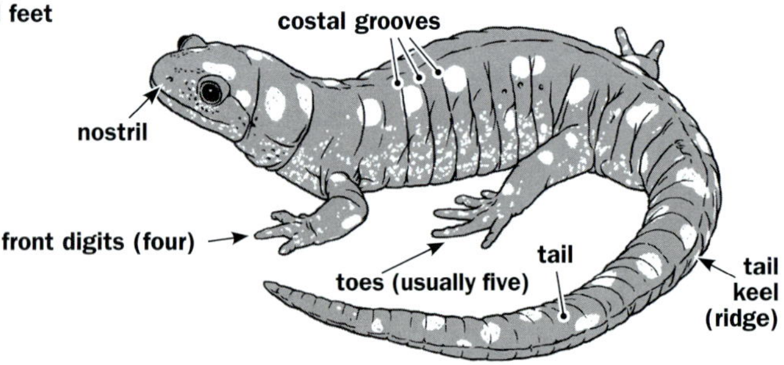

Adult Toad Features

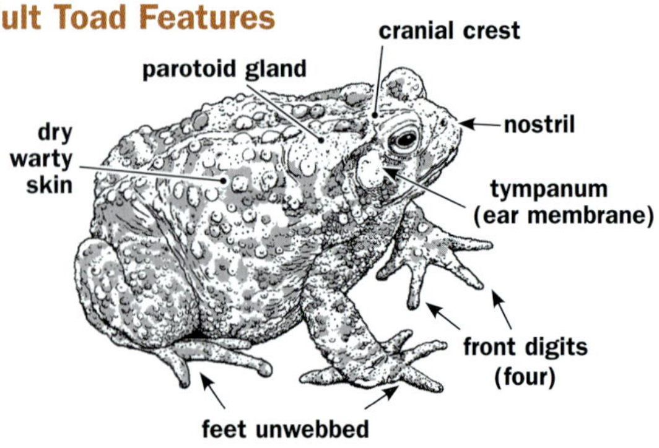

Amphibians evolved from fish about 350 million years ago, and modern amphibians appeared about 200 million years ago. Today's amphibians either remain in water all their lives, live on land at least part of their lives, or live on land all their lives. But even the latter have not escaped the need for a moist environment.

Amphibians are characterized by their moist skin, which is susceptible to desiccation because there is no protective covering of scales, feathers or hair to help retain moisture. An amphibian's skin might not look impressive, but it's actually very specialized. The skin is permeable and absorbs water. It's a breathing organ, even though most amphibians have lungs. Amphibian skin is loaded with tiny glands, which produce substances of varying toxicity or at least give a bitter taste to a would-be predator. The skin of some amphibians displays bright warning colors to deter potential predators, while the skin of others provides camouflage. Some amphibians can change skin color to match their environment.

Two orders of amphibians inhabit our three northern states: *Caudata*—the salamanders, and *Anura*—the frogs and toads.

Amphibian Biology 101

Amphibians, as might be expected by their moist skin and jelly-like eggs, need moist conditions for breeding, usually in the form of a pond. Male amphibians have no copulatory organs, so fertilization of eggs is either indirectly internal, as in most salamanders, or external, as in frogs and toads.

Frog and Toad Reproduction

Some species of frogs breed in early spring, often using the same vernal ponds as early breeding salamanders. Other species of anurans breed later in the spring and still others breed during summer. The frogs that procrastinate until summer to breed don't need to hurry. They breed in permanent waters, and the tadpoles can overwinter before metamorphosing the next year or even two years later.

Male frogs and toads vocalize to attract a female. Each species has its own distinct call, which is made with a single vocal pouch or a pair of vocal pouches, depending on the species. Males of some species battle for territory by pushing and wrestling. A winning male becomes the resident of a prime calling site. The losers of these battles look for

The male Spring Peeper "peeps" by inflating his single vocal pouch. He hopes to attract a mate.

Life cycle of the Wood Frog
1. Vocalizing male
2. Mating pair in amplexus
3. Egg mass
4. Tadpole
5. Newly metamorphosed frog

a different location or become satellite males. Satellite males hang around, keeping a low profile, until the resident male successfully attracts a female and moves off with her to breed. Then one of the satellite males will take his place. A satellite male might also intercept a female before she gets to the resident male. The male climbs on the female's back and clasps her behind her front legs. This embrace is called amplexus. To better maintain his grip on the female, the male has thickened thumbs during the breeding season. As she lays her eggs, he fertilizes them. Males are anxious to breed and sometimes grab another male by mistake. The grabbed male sounds a release call to point out the error. Eggs are laid in various sized clusters or strings, depending on species.

The eggs hatch into tadpoles. A tadpole has a large combined head and body with a vertically flattened tail having upper and lower fins for swimming. Newly hatched tadpoles have visible gills, but these are soon covered by the opercular fold. The hidden gills still absorb oxygen as water moves through the mouth, over the gills, and out the spiracle, which is a tube located on the left side of the tadpole. Tadpoles have large eyes and scraping mouth parts. They are mainly herbivorous. As tadpoles grow, the hind legs appear first, while the front legs are hidden by the opercular fold. When the front legs appear, the tadpole is close to changing to the adult form. Its tail fin disappears and the muscular portion of the tail becomes smaller. The tadpole changes from herbivore to carnivore, and it changes from breathing with gills to lungs. When this happens, the frog needs to be able to climb out of the water or it may drown. It leaves the water before the tail is completely absorbed.

Frog Life-Cycle

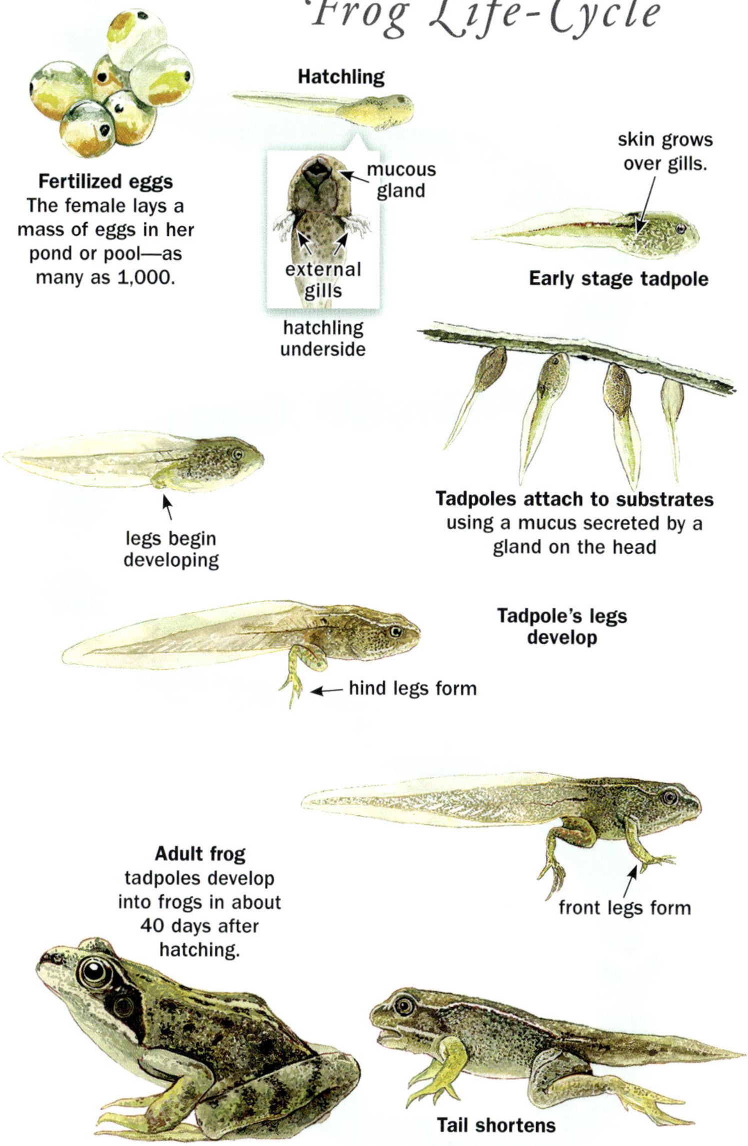

The Wood Frog Life Cycle

Wood Frogs, along with spring peepers and chorus frogs, are the earliest of our anurans to breed. Ice may still be on the pond when warm spring rains spur the Wood Frogs to vernal pools, bogs or other small woodland wetlands.

Male Wood Frogs don't fight for territories, but they float at the

Sticky Feet

Tree frogs have remarkable, gravity-defying toe pads that can attach to smooth, rough, dry and wet surfaces. These pads are covered in minute hexagonal structures: sticky mucus gets secreted between them and acts as an adhesive.

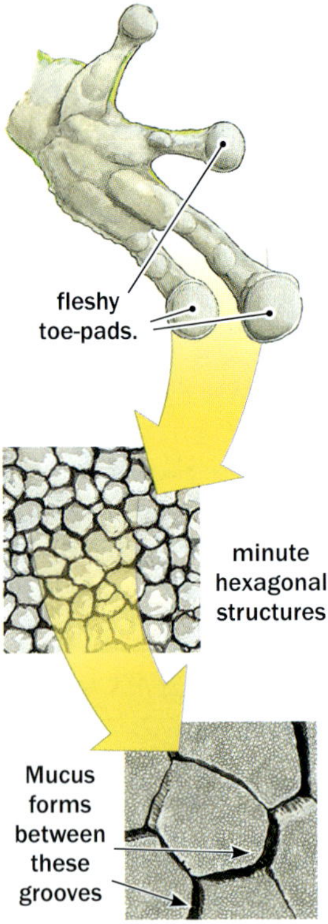

surface of the water and vocalize for females. A pair of vocal pouches, one above each front leg, are rapidly inflated and deflated in a staccato, duck-like quacking call. A male seized mistakenly by another male will give out a chirp, which is the release call. A female ready to breed will allow a male to embrace her. She then swims, carrying the male on her back, to a site for laying her eggs. The male fertilizes the eggs as the female expels them where they will stick to a submerged twig or aquatic stem. The hundreds of eggs quickly absorb water and swell to a globular mass the size of a tennis ball.

Sometimes many females choose the same small area in a pond for their eggs. Possibly communal nesting offers protection from predators, because so many egg clutches in one place couldn't all be eaten before some hatch. More likely, the dark gray or black eggs, warm by absorbing sunlight, and the combined effect of many clutches actually help warm the shallow pond water in that location.

The tadpoles usually hatch in two to three weeks, but the incubation time could be quicker or longer, depending on temperature. The tadpoles stay with the disintegrating egg mass for a few days before dispersing. The tadpoles have scraping mouthparts and are primarily herbivorous, eating algae and other plant matter, but they may also feed on other amphibian eggs or larvae. Many tadpoles are racing the season, because their pond will dry. Some years the tadpoles don't have enough time and die when the pond dries. If the pond retains water long enough, the tadpoles develop front legs in addition to the rear legs, lose their internal gills and tail fin, and the tail begins to be absorbed. The tiny frog crawls out of the water.

The new batch of Wood Frogs leaves the area around the pond and disperses into the woods. The little Wood Frogs, like the adults, are now strictly carnivorous, feeding on insects and other invertebrates. Males reach maturity in one or two years, while females require two or three years.

Salamander Reproduction

Most salamanders in our area emerge from hibernation with the first good rain during late March or early April. The rain, ideally an electrical storm lasting into the night, acts as a wakeup call to the salamanders and synchronizes their migration to a breeding pond.

Some salamanders require fishless ponds, otherwise the adult salamanders, the eggs, and the salamander larvae end up as fish food. A pond shallow enough to freeze out fish, or a vernal pond, which is a temporary spring pond formed by melting snow and spring rains, is a necessity for these salamanders. Vernal ponds usually dry by mid or late summer, which is a good reason for early breeding, to give the larvae more time to metamorphose and get out.

The indirect internal fertilization of salamander eggs is preceded by courtship, which is primarily rubbing and nudging between the male and female. The male deposits spermatophores, which are jelly-like structures capped by a packet of sperm. The female picks up spermatophores or just the sperm packet with the lips of her cloaca, which is the chamber used for reproduction or elimination of wastes. The sperm is stored in the spermatheca, located within the female's cloaca, until she lays the eggs, which depending on the species could be a few minutes, a couple days, or several months. When the eggs pass through the cloaca, they are fertilized. Salamander eggs laid in water hatch into carnivorous larvae. Salamander larvae have obvious gills, elongated bodies, and have similar sized front and rear legs. A larva metamorphosing into a terrestrial adult will lose its gills and the fin from its tail. Some salamanders lay eggs on land, skipping the aquatic larvae stage.

Spotted Salamander Life Cycle

Spotted Salamanders are very early breeders, migrating during a rainy night in late March or April from their wooded upland hibernation sites. They are often a couple days behind the Blue-spotted Salamanders, which hibernate closer to the pond. Spotted Salamanders use vernal pools or other shallow, fishless waters for breeding. The males arrive first and hang out in groups called congresses. The females usually arrive a few hours later, but could be delayed for days if the

Spotted Salamander

Life cycle of the Spotted Salamander
1. Adult
2. Female with egg mass
3. Newly-hatched larvae in "jelly"
4. Fully developed larva
5. Newly metamorphosed salamander

rain ends quickly.

A male and female perform a dance-like courtship on the bottom of the pond, as they circle each other and nudge the base of each other's tail. Eventually the male will swim off with the female following. He then deposits a spermatophore, a jelly-like blob capped by a sperm packet, onto a leaf or twig on the pond bottom. The female moves over the sperm packet and takes it into her cloaca. Males deposit several spermatophores, and females may pick up more than one, not necessarily from the same male. She retains the sperm for a day or two. Then the sperm will fertilize the eggs as she lays them.

The female finds a stem or thin branch near the surface and lays her eggs on it. The eggs and a jelly-like substance swell to the size of a tennis ball, leaving the eggs embedded within the firm clear or milky white material. The eggs may turn green from algae in a symbiotic relationship, where the algae gets carbon dioxide from the egg and the egg gets oxygen generated by the algae. It takes the eggs about three weeks to two months to hatch, depending on water temperature. The temperature of the water also affects development of the carnivorous larvae, which need at least a couple months to develop. They usually metamorphose in late summer. A drying pond will stimulate faster development. If necessary, the larvae can overwinter in permanent bodies of water. Metamorphosis to the terrestrial form means the tail fin and gills are lost, and breathing is achieved with lungs. Newly metamorphosed salamanders stay near the pond for a while, hiding under leaves and logs. On a rainy September night they migrate to the wooded uplands. Males are mature after two or three years and females after three to five years.

Meet the Reptiles

Reptiles evolved from amphibians about 335 million years ago with innovations to survive in a drier environment. They have dry scaly skin which retains moisture and provides protection. Scales can be in various forms, from tiny granular scales to large shields. The structure of reptile scales can offer protection from predators, as in the outer covering of a turtle's shell. The color of a reptile's skin often provides camouflage. Other reptiles have warning colors. Some lizards have a brightly colored tail which acts to distract a predator. The predator grabs the tail, which promptly breaks off, and the lizard escapes.

The front foot of a reptile has five toes instead of four, and most toes have claws.

Reptiles abandoned the idea of reproducing with a gilled larval stage in a pond. Instead they have internal fertilization, which allows reptile eggs to have moisture-retaining leathery or calcareous shells. The eggs can incubate in the ground or in a moist rotting log. Some reptiles have further evolved to bear live young.

Because reptiles are ectothermic, much of their behavior involves maintaining an optimum body temperature. Turtles are especially noticeable, as they may spend hours basking in the sun. Lizards and snakes warm faster in the sun. They often move between sun and shade or warm themselves underneath a flat rock when the upper surface is too hot. Seasonal temperature changes influence when some species are nocturnal or diurnal and when they are active or dormant.

Two orders of reptiles are found in our northern three states: *Testudines*—the turtles, and *Squamata*—the lizards and snakes.

Reptile Biology 101

The reptile egg was a significant advance over jelly-like amphibian eggs, most of which require a pond. All reptiles have internal fertilization, which is a prerequisite to covering the egg with a shell. Inside the calcareous or leathery shell, a membrane called the amnion surrounds the embryo. The amnion helps retain moisture and insulate the embryo against temperature changes. Two other membranes, the chorion and allantois, contain blood vessels and are involved with oxygen transfer through the permeable shell to the developing embryo. The allantois also collects the embryo's wastes. A yolk sac supplies nutrients to the embryo.

Most reptiles are oviparous, meaning they lay eggs. All turtles are oviparous. All our lizards and about half of our snakes are oviparous. The other snakes of our region bear live young. But unlike most mammals which are viviparous, the snake embryos receive no direct nourishment

Reptilian External Features

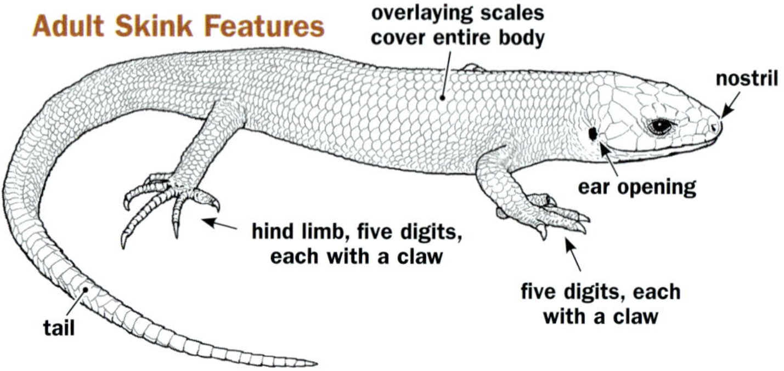

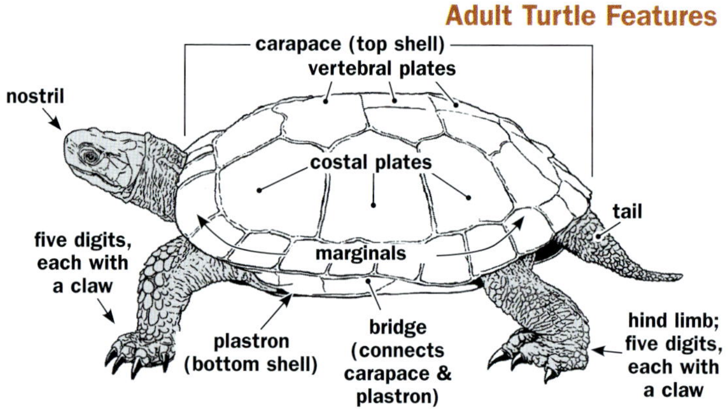

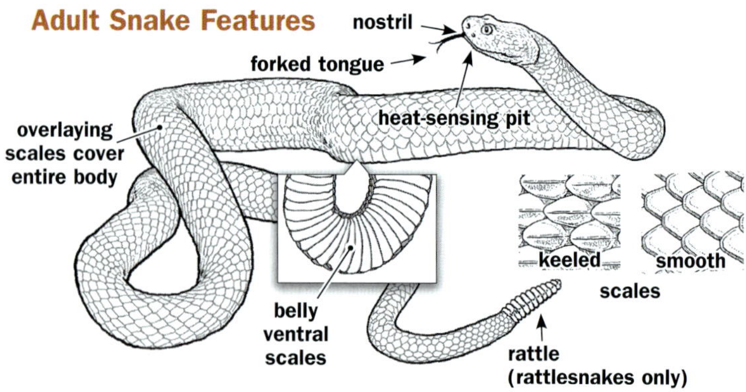

from the mother. Instead, each embryo has its own yolk sac. Because of this difference, live-bearing snakes are said to be ovoviviparous. When born, each juvenile snake emerges from its enclosing membrane.

There are actually different kinds of reptile eggs. One kind is hard-shelled and brittle like the eggs of the Eastern Musk Turtle or the Eastern Spiny Softshell. These eggs retain moisture better which helps when the musk turtle lays her eggs here and there without even burying them, or the female softshell nests in a sand bar. Other eggs have soft leathery shells, some of which can expand as they develop and sometimes become distorted or bulged out. Examples of this are Prairie Ringnecked Snake eggs. Some species, like the Eastern Hog-nosed Snake, have eggshells thin enough to see the snake inside as they begin hatching. Many clutches of snake eggs adhere together. I assume this is to protect them from being devoured by another snake, as the eggs are hidden but abandoned by the mother.

Most reptiles lay eggs with soft, leathery shells (like the Prairie Ringnecked Snake, top photo) but a few have eggs with a hard exteriors, similar to birds' eggs (like the Eastern Musk Turtle, bottom photo).

Sometimes reptile eggs are protected by the mother, as is those of the Common Five-lined Skink.

Turtle Reproduction

Many of our turtles mate in both fall and spring. Male turtles of some aquatic species have very long front claws which are fluttered in front of the female's face and stroke the sides of her head. Courting male box turtles nip at the female's shell or feet.

A male turtle climbs the female's back and holds her shell with his claws. His longer tail twists under hers until their cloacas are in contact, and his penis can be inserted.

Turtles nest from late May into early July. Some species nest on a nearby sandbar while others travel some distance from water. The female may dig in several places before settling on the site for her eggs. In most species, the female digs a flask-shaped hole with her back feet and sometimes voids water to moisten the soil. The eggs are deposited into the nest. The female uses her back feet to cover the nest and may even crawl over it to complete the task.

Most turtle nests are dug up and destroyed by skunks, raccoons, foxes, or other predators. Sometimes the predator doesn't even wait for the turtle to finish. One night I observed a Blanding's Turtle nesting and I moved on. Returning later, I found a badger, the upside-down turtle, and an empty nest.

The turtles hatch in late summer. The egg is sliced open by a sharp, pointed, tooth-like structure on the turtle's snout called a caruncle. The hatchlings dig together to the surface. But the turtles of many nests overwinter there and don't emerge until spring.

Painted Turtle Life Cycle

The male Painted Turtle swims backwards in front of the female and strokes the sides of her head and neck with his long front claws. If she is receptive she strokes his front legs. The male swims around behind and then over the female and holds onto her carapace. He twists his tail under hers and they sink to the bottom of the pond.

In late May or June the female crawls ashore and looks for a nesting site. She uses her back feet to dig the flask-shaped hole, often wetting the soil to soften it for digging. She lays 3 to 20, usually 7 to 9, elliptical eggs. She scrapes the excavated soil back into the nest and smoothes it over. She returns to the pond. Most likely, her eggs will be devoured by a predator the first night or two. If they beat the odds, they hatch in 50 to 80 days.

The sex of the hatchlings was previously determined by their incubation temperature at a certain critical time. If it was 75 to 80 degrees, all hatchlings are males. If it was 86 to 90 degrees, all are females. The hatchlings may stay in the nest until spring. Although subjected to freezing temperatures, the little turtles are protected by a natural antifreeze. But they may not survive extremely cold weather combined with lack of snow cover. In spring, usually after a soaking rain softens the soil, the little turtles crawl out and head for the pond. Those lucky enough to elude predators reach adult size in 4 to 6 years.

Life cycle of the Painted Turtle

1. Female looks for egg-laying site
2. Laying eggs in soft soil
3. Hatching from egg
4. Hatchling turtle

Lizard Reproduction

Most lizards are territorial. Breeding colors and behaviors, such as head-bobbing, help establish territories and identify potential mates or intruding males. When a male finds a receptive female, he grasps her neck in his jaws, and she raises her tail slightly. The male will move the base of his tail against hers. Male lizards are equipped with paired copulatory organs called hemipenes. The male inserts one hemipenis into the female during mating. Most lizards lay eggs, and some tend the eggs until they hatch.

Common Five-lined Skink Life Cycle

In spring, the male Common Five-lined Skink develops a bright red-orange coloration around his nose and mouth. He establishes and defends a territory about 30 yards in diameter. Female and juvenile skinks lack the orange-red coloration. Also, females have a bluish tail, and juveniles have a bright metallic blue tail. They are allowed in the male's territory. Intruding males are driven away. The male holds the female's neck in his jaws, twists his tail under hers, and inserts one hemipenis into her cloaca. The female becomes noticeably distended with eggs during early summer. She lays 5 to 18 eggs in a moist place under a log or in a burrow. She stays to tend the eggs. She basks in the sun and then warms the eggs by curling around them. By urinating in the nest, she adds humidity. She turns the eggs. If an egg spoils, she eats it. She guards the eggs from small predators. In 24 to 55 days the eggs hatch. The young, with their vivid stripes and metallic blue tails will be

Life cycle of the Five-lined Skink

1. Female in burrow
2. Eggs are guarded by the female
3. Emerging from the egg
4. Juvenile has a bright blue tail

guarded by the mother for a day or two until they disperse.

Snake Reproduction

Most snakes breed in spring or fall when they are gathered in or around their hibernaculum. The male snake can also track a female using his tongue and Jacobson's organ to sense her pheromones. Sexual recognition is by scent, since there are no obvious differences in appearance. Some male snakes engage in a form of combat, in which they wrap around each other while trying to raise above the rival and then push him down. Most male snakes will move with a jerking motion when along side a potential mate. Sometimes a rippling motion travels along his body from head to tail. Like the lizards, the male snake usually grasps the female's neck in his mouth. The snakes wrap their bodies around each other and the male uses a hemipenis for internal fertilization. Depending on species, the female lays eggs in late June or July, or bears live young in August or September. Snakes that hatch from eggs have a special egg tooth to slice open the leathery shell.

Milksnake Life Cycle

Milksnakes mate after emerging from hibernation or later in the spring. The male puts his head on top of the female and wraps his body against hers. He might hold her neck in his mouth. When their cloacas are together he inserts a hemipenis. In early July the female Milksnake lays 5-24 eggs in a hidden, warm, moist cavity under a rock or rotting log. The eggs stick together in a cluster, possibly to prevent another snake from eating them. The eggs get no more protection from the female,

Life cycle of the Milksnake

1. Adult Milksnake
2. Female with her egg cluster
3. Juvenile cuts shell with egg tooth
4. Hatching from the egg

because she abandons them. They hatch in 6 to 9 weeks. The juvenile Milksnakes are more colorful than the adults, especially when, a few days after hatching, they shed for the first time. They reach maturity in their third or fourth year.

Finding and Observing

Decades ago I was looking for my first Western Foxsnake. It was mid-May, and I was on an abandoned railroad embankment that ran through wetlands. I heard a buzzing sound. It could have been an insect, but I looked closer. I spotted the source: a vibrating foxsnake tail. My eye moved from the tail up the body, and there it was: a Western Foxsnake half coiled and basically in plain sight. Its light brown body with dark brown blotches was perfect camouflage for the heavy gravel substrate. Later that day I found another foxsnake blending into a similar background.

There are a number of lessons here. I knew what I was looking for, what it looked like, and that it might buzz its tail. I knew the foxsnake's habitat, and that mid-May would be a good time to find one warming itself in the sun. Also, in mid-May the area isn't overgrown with vegetation. After finding the second one, I was convinced that you can develop an eye for spotting the color and pattern of a species.

A cold spring pond is as unexplored as the wildest woods; get out and explore for herps.

The lesson I didn't learn then, was that once you grab the herp, you can no longer observe it in the wild. More recently, I saw a Wood Turtle walking along the sandy bottom of a shallow, clear stream. It stopped as it seemed to see me. I didn't move. The turtle scooped sand with its front feet and flipped it up onto its carapace. Then it moved closer, poked its head out of the water, and looked at me. Did it feel safer – better camouflaged – with the sand on its back? I've also seen Wood Turtles on shore flip sand up on the carapace. Wood Turtles have concentric ridges on their carapace scutes. Did the ridges evolve to help hold sand grains for camouflage? More observation is needed.

The first thing to remember about finding amphibians and reptiles is to be alert for them, whether you're driving down the highway or hiking through the woods. When hiking, move slow and look close. Many herps will just freeze, relying on cryptic coloration while waiting for you to pass by. Binoculars are useful for observing and identifying turtles out on logs and to spot less obvious turtles, frogs, and watersnakes along the shore.

One of the best ways to find amphibians and reptiles, especially lizards and snakes, is to look under objects such as flat rocks, boards, sheets of tin and logs. Lift the object from the far side and pull it towards you, as a precaution against venomous snakes or wasps. If you lift up a flat rock that's in the sun, feel the underside. Is it slightly warm? Is it too hot for a reptile? If it's too hot, lift thicker rocks, or next time try earlier in the day.

When you lift objects, always replace them as you found them. That is home for many creatures. The rock may cover burrow entrances for invertebrates and possibly herps as well. It's inexcusable to ruin that microhabitat, which may have taken years to develop. Always be careful to avoid damage to the environment.

Search for herps at night with a good flashlight or headlamp. Some snakes and lots of amphibians will be out at night. Check around porch lights, campground lights, or lit signs for opportunistic anurans devouring attracted insects.

Frogs and toads are best found at their breeding ponds. Knowing what the frog looks like and what it sounds like will make it easier to find. It helps to know how big a frog is or if it could be calling from up on a branch or from under the water. Calls can be learned from recordings listed in the back of this book.

Night is the best time to approach and observe frogs. A warm, wet night is best, because the frogs tend to come out and perch in more obvious places. The frog chorus should be enthusiastic and quick to resume if disturbed. Some frogs can be observed from the shore, but you'll see more if you enter the pond. Rubber boots help, but chest waders allow you to kneel down in one or two feet of water and be close to the frogs.

The best way to observe frogs calling is to wade into their midst, turn off your flashlight, and wait for them to resume calling. When the chorus is going strong again, turn on your light and search. If you're looking for chorus frogs or Spring Peepers, don't be fooled by their decibel level. These frogs are barely over an inch long and are cryptically colored. You'll probably spot the bright vocal pouch first. Besides looking for individual vocalizing male frogs, look all around you. Females don't call. Frogs in amplexus aren't calling. Right next to you might be a species not yet breeding. A salamander or snake might be there also.

Amphibians and reptiles are amazing creatures that have been around for a very long time. I'm sure there are still many things about them we don't know. I would urge people to resist the temptation to grab, and instead, see what can be learned from observing amphibians and reptiles in the wild.

Seasonal Guide to Herps in Minnesota, Wisconsin & Michigan*

Winter: Good luck. If you can find a small place at the edge of a lake that's kept free of ice by a spring, you might find newts. You might catch a mudpuppy while ice fishing with worms.

Early April: Warm rain would spur the night migration of mole salamanders.

Mid April: Look for mole salamanders on a rainy night if they didn't run earlier. Listen especially for Wood Frogs, because their breeding season is short. Chorus frogs and Spring Peepers also start calling.

Late April: Mole salamanders may still migrate in the north. Northern Leopard Frogs join the choruses of Wood Frogs, chorus frogs and Spring Peepers. Look for basking Painted, Spotted, and Blanding's Turtles. Look for gartersnakes at south-facing den sites.

Early May: Spring Peepers and Northern Leopard Frogs are still calling. Toads calling. Hatchling Painted Turtles that overwintered in the nest are on roads. Even at 55 mph you can see the lumbering gait compared to a still rock. Adult turtles are basking in large numbers.

Mid May: Treefrogs are calling. Most snakes are out. Check under logs surrounding vernal ponds for salamanders.

Late May: Treefrogs calling. Four-toed Salamanders guarding eggs. Snakes migrating.

June: Treefrogs and Green Frogs calling. Turtles nesting. Look for Red-backed Salamanders foraging in the woods on wet nights. Look for mudpuppy eggs attached underneath large submerged rocks.

July: American Toad juveniles on pond shores. Red-backed Salamanders guarding eggs.

August: Look for juvenile treefrogs on tall grass or milkweeds surrounding ponds or marshes.

September: Newly metamorphosed mole salamanders migrating on rainy nights. Turtles hatching. Snake eggs hatching or being born alive. Snakes migrating across roads.

October: Snakes at dens.

*Spring may come early or late and cold spells can delay events. The order of events is a better guideline than the dates.

Catching and Studying

Amphibians and reptiles are studied by herpetologists who want to learn more about them and their role in the environment. One species in a designated area may be studied over time to estimate how many are there and if the population is increasing or declining. Sometimes a specific area is studied to learn what species are found there, and then which places are most important to protect. Disappearing and deformed frogs have been the focus of many studies that are still ongoing.

This Blanding's Turtle is numbered for a Minnesota tracking study.

Frog and toad population surveys are done three times a year during designated periods to include the breeding season of all anurans. The surveyor travels the same route and stops at the same breeding ponds. The species heard calling are recorded, along with an estimate of calling intensity and temperature data. Over the years, population trends can be demonstrated.

Most amphibians and reptiles have to be captured for study. A team of people may simply search an area and grab every amphibian and reptile they can find. Nets are used to capture aquatic herps, while snake hooks and snake bags are the most common terrestrial tools. In preparation for a survey, boards may be laid out at intervals in likely places.

Dan Nedrelo sets a turtle trap for a Wisconsin Department of Natural Resources study.

Weeks later they are checked for whatever herps might be underneath. Drift fences are often used in surveys. These can be made from various materials, including rolls of thin aluminum, hardware cloth, chicken wire and the black plastic used to prevent erosion around construction sites. The drift fence is stretched out to block the path an amphibian or reptile is likely to travel while migrating or searching for a nesting site. Sometimes funnel traps are placed along the fence, or pit traps are dug at each end. The pit trap may be a jar or can with the top at ground level. A board can be placed over it, so a lizard dashing under the board will fall into the container.

Top: A Massasauga is implanted with a tiny transponder. Bottom: The same snake can be identified with the use of a scanner when recaptured.

Large hoop nets are used with bait to capture turtles. The nets are partly above water so the turtles can breathe.

Snakes may be caught in funnel traps or captured by hand, especially around den sites. Depending on the study, captured herps are usually marked in some way so they can be identified if recaptured later. The usual method for marking turtles is to file notches in the marginal scutes. One method of marking snakes is to insert a tiny passive implant transponder tag (PIT tag) underneath the skin. When the snake is recaptured, it can be identified with a device that scans and reads the tag through the skin. Sometimes an animal is equipped with a transmitter so it can be relocated by radio-telemetry.

Studying amphibians and reptiles doesn't have to be an official project. Anyone can gather data. It starts with a notebook. Whenever you find an amphibian or reptile, record the species or subspecies, sex (if known), approximate age, date, location, time of day, weather conditions and other information, such as AOR (alive on road), DOR (dead on road), basking on rock, under a log, etc. I started keeping records on herps when I was in eighth grade and I refer to some of that data nearly every year.

Conservation

Amphibians and reptiles are under siege. Most species are diminishing in numbers due to a multitude of reasons, especially environmental degradation, and, unless I'm overlooking something, all are ultimately the fault of man.

All too often, reptiles are deliberately killed. Of course snakes are killed by people who are ignorant, afraid, and/or just plain ruthless. But even turtles are killed. There are actually people who deliberately drive over turtles or shoot them for target practice. A well-known country music singer drives over box turtles with his pickup truck and calls it turtle popping.

Some people remove reptiles from the wild. A tourist finds a Wood Turtle or a Spotted Turtle, both of which are protected, and takes it home for a pet. A few reptile enthusiasts, who know better, might take one as well. These individual losses to the population add up. It takes years to replace an adult turtle, and the mortality of eggs and young is very high. But far worse is the problem with poachers taking them for the pet trade. Wood Turtles have been especially hard hit as poachers take every one they can find, literally decimating the population along some stretches of river. Wood Turtle population surveys, carried on for years where the turtles were plentiful, have been terminated because one year the turtles were missing. Spotted Turtle and Box Turtle populations have also been hurt. A few poachers have been busted for selling the turtles in other states and to

Female Snapping Turtles are especially vulnerable when searching for nesting sites.

Wetlands continue to be filled, dooming the reptiles and amphibians that need them.

China where the turtles are eaten.

Disappearing habitat is a serious problem for amphibians and reptiles. Development is constantly destroying habitat. Displaced animals can't simply move somewhere else, because somewhere else is usually already occupied. A habitat is limited in how many of any species can live there. That limit is called the carrying capacity. The limiting factor may be food, shelter, or something else, but when the carrying capacity is reached, there is no room for more.

Green Herons are a natural hazard in the wild.

Wetlands are often destroyed by development. The doomed resident amphibians and reptiles don't care that five miles down the road a big hole is being dug and filled with water as a mitigation site. Vernal ponds often get filled because they are dry part of the year, but that is why some amphibians breed there. The addition of fish to a small pond is destructive, because they devour the salamander larvae and tadpoles.

Poorly placed roads, such as through wetlands or between forest and pond, can result in high mortality of herps during migrations.

Clear-cuts doom salamanders. Even selective cutting makes the forest drier and less hospitable for amphibians.

More sinister assaults on amphibians and reptiles involve the widespread deterioration of the environment by polluting chemicals. Amphibians are especially vulnerable because their skin and unshelled eggs are exposed to water, soil, and sun. Studies have shown that increased ultraviolet-B radiation, due to the thinning ozone layer, can kill amphibian eggs.

Frogs received a lot of publicity when large numbers were found deformed in 1995. Frogs have been found with extra legs, misplaced legs, missing legs, missing eyes, and even more grotesque deformities. Some deformities are caused when the larva of a parasitic flatworm forms a cyst in the developing leg of a tadpole. But many scientists believe other factors cause deformities, as well.

Frog populations have declined or disappeared in some places. If a cause, such as disease, fungus, or parasite is suspected, the real culprit may be much more complex. The frog's immune system may have been weakened first by a chemical pollutant, making the frog more susceptible.

Acid rain, PCBs, mercury, pesticides, and herbicides are all threats to amphibians. Pesticides and herbicides can contaminate breeding ponds, either from drifting spray, or from rainwater runoff.

Amphibian deformities may be the result of agricultural runoff or some other environmental contaminant.

Studies by Hayes showed that a widely used herbicide caused sexual abnormalities which feminized male Northern Leopard Frogs. When tadpoles developed in water containing atrazine at 0.1 parts per billion (the EPA's limit for drinking water is 30 times higher!) the resulting hermaphroditic frogs had both ovaries and testes, or they had egg cells inside the testes. Further studies by Hayes have found that combinations of commonly used pesticides, herbicides, and fungicides have more severe effects than would result from the same concentration of only one.

A study by Relyea found that a commonly used herbicide, Roundup, used at the recommended amount, killed frogs and toads on land and tadpoles in the water. On land, 79% of the frogs and toads died within one day. In the water, 98% of the tadpoles died.

The anurans used in the study were the American Toad, Gray Treefrog, and Northern Leopard Frog. I first wrote these words fourteen years ago. Roundup is still popular and is still killing frogs and tadpoles.

Many of our amphibian and reptile populations are declining. Federal regulations protecting wetlands have been weakened and are still under attack. Quality habitats have been drained, filled, polluted, or otherwise destroyed. To aid in the destruction of wetlands, the state of Wisconsin removed the Blanding's Turtle from the Threatened Species list a few years ago so more wetlands could be filled and developed.

Climate change is causing problems for all life on Earth. It is getting warmer and warmer. Man is to blame. Some politicians dispute that. They say it's still debatable. They delete the words "climate change" from state websites and ban state employees from using those very words.

A lot of people don't like politics, but politics will decide what habitats and species will still be around years from now. Anyone who cares about amphibians, reptiles, and other wildlife needs to start paying attention. Some politicians care about environmental protections.

Others would sell off parks and forests to balance the budget, or drive over a vernal pond in August to show it's not much of a wetland. Know who is who. Don't listen to them; check their environmental voting records. Support organizations that protect habitat. Future generations might want to hear frogs in a pond or watch turtles on a log. What a shame if those pleasures were forever lost.

How to use this Field Guide

Amphibians and Reptiles of Minnesota, Wisconsin & Michigan is designed to make field identification easier for you. Through the use of color photos, arrows pointing to field marks, maps, habitats and color tabs, we have made a handy, compact and easy to use guide. Also, by limiting the amphibians and reptiles to those found in one geographic area, we have eliminated the need to wade through several hundred species, many of which would never be found here. Included is every regularly occurring herp in the three state region of the Western Great Lakes.

Amphibian and Reptile Names Like other organisms, herps are given both common and scientific names. The common names are the English names most amateur naturalists use, while the scientific or Latin names tend to be the spoken word of herpetologists. We capitalize species names but use lower case for groups (e.g. *B*utler's *G*artersnakes are actually a type of a *g*artersnake). If there is a distinct subspecies in our area, we list it. For example, there are two subspecies of Painted Turtle in our three-state region (Western Painted Turtle and Midland Painted Turtle) and we cover them both.

Taxonomy is a fluid science; names are always in a state of flux. We've decided to use the list recognized by several professional organizations. Go to *www.herplit.com/SSAR/circulars/HC29/Crother.html* and click on "*PDF of Book*" or Google "*herplit list north america.*"

Photos Nearly all photos are by the author. Many were taken in the wild but if the subjects were temporarily captive animals, Allen was careful to keep them in their natural habitat. We've attempted to show as many behaviors and stages of the species life cycle as we could. The symbol used for males is ♂ and the female symbol is ♀

Habitat Preferred habitat is where courtship, mating and egg laying takes place, and the larva develops and emerges. Your best bet to find the species will probably be near such habitat—but not exclusively.

Official State Status If a species has official state status, we list it on the main photo: Endangered, Threatened, Special Concern or SGCN (Species of Greatest Conservation Need). Details on next page.

Nature Notes These are fascinating bits of natural history that bring one a more complete understanding of that species. Unique defense behavior, hunting techniques and diet preferences are just some of the topics touched on.

Range Map A map of Minnesota, Wisconsin and Michigan shows the range of that species in pink. This is the range as we know it from current population data but some amphibian and reptile populations are poorly understood or changing.

Species Range

Species Text **Description** covers the best distinguishing field characteristics. If males and females are different (sexual dimorphism) then these differences are described. Measurements given are the average body length in inches. For skinks, the measurement is snout-to-vent length because the tail can break off. Salamander measurements, though, are snout to end of tail. Turtle measurements are the straight-line carapace length.

Under **Range Notes** we note unique aspects to a species range that is not shown by the map, and also the general North American range.

Courtship, Mating & Eggs covers exactly that—except for species that bear live young (several species of snakes). Reptiles and amphibians have unique courtship rituals that are discussed here. Salamanders and frogs progress through several distinct life stages and these are also covered.

Diet is only covered for snakes and turtles, who can have very specific prey preferences.

Glossary Check out the glossary for easy-to-understand meanings of some tricky terms.

Titles of Interest This list of recommended reading and resources includes our favorite titles for delving deeper into the fascinating world of herps. Also listed are books on close-up nature photography. **Additional References** lists other pertinent research papers and articles. You may also want to pick up a CD of frog and toad calls to better equip yourself for field identification. Several sources are listed here.

Enjoy *Amphibians and Reptiles of Minnesota, Wisconsin and Michigan.* Take it with on hikes. Stuff it in your pack. Use it. But most importantly, have fun getting to know our fascinating northern herps.

Blue buttons estimate the relative abundance of that species

Habitats where this species typically occurs

Codes for special status in MN, WI or MI (see below)

Photos showing other views, habitat or details can be found on the right-hand page

Nature Notes **are natural history tidbits about that species**

Range maps show known distribution of that species

Color tabs quickly get you to the group of herps you are looking for

Family name is listed on the lower left hand page and genus is found on the right page

Quick Color Tabs on the page edges and back cover help you get quickly to the right section of the guide.

salamanders	toads	frogs	turtles	lizards	snakes

Official State Status for vulnerable species

MN [Minnesota]
WI [Wisconsin]
MI [Michigan]

Endangered [In danger of extinctinction in the state]

Threatened [Likely to become endangered in the state if no action taken]

Special Concern [Rare species, vulnerable species, or species with very specific habitat requirements]

SGCN [Not an official status such as those above but Species of Greatest Conservation Need; MN only]

Order Caudata: Salamanders

Salamanders are amphibians that have tails as adults. They have elongated bodies, and most have four legs of relatively equal size. There is no tympanum. Salamanders breathe with one or more of the following organs: gills, lungs, or skin. The approximately 410 species known throughout the world are grouped into 10 families. In our three states, we have twelve species, but they represent the diversity of four families.

Family Proteidae – Mudpuppies and Waterdogs

This family has only six species worldwide. All are neotenic, keeping their gills and remaining totally aquatic through adulthood. We have a single species: the Mudpuppy.

Family Salamandridae – Newts

About 55 species of newts occur worldwide. Newts are not slimy; they have drier, rougher skin than other salamanders. They are primarily aquatic with tail fins but the adults don't have gills. One species lives in our area: Eastern Newt.

Family Ambystomatidae – Mole Salamanders

There are at least 30 species of mole salamanders, which are found only in North America. These are stout medium-sized salamanders that spend most of there lives underground or at least underneath logs or other shelter. Adults have four toes on each front foot and five on each rear foot. Mole salamanders have distinct vertical creases on their sides called costal grooves that can be useful in identification. Six species are found in our area: Blue-spotted Salamander, Spotted Salamander, Western Tiger Salamander, Marbled Salamander, Small-mouthed Salamander and Tiger Salamander.

Family Plethodontidae – Lungless Salamanders

This is a large family with about 266 species worldwide. These salamanders actually are lungless, breathing only through their skin and the lining of the mouth. Dependent on breathing through the skin, they are necessarily small. Members of the Plethodontidae have a unique feature called a nasolabial groove. This tiny groove runs from the nostril to the mouth and is believed to enhance the sense of smell. In some species the groove ends in a tiny projection called a cirrus that extends just below the mouth line. Cirri are sometimes more prominent in males. Four species of lungless salamanders are found in our three states: Northern Dusky Salamander, Southern Two-lined Salamander, Four-toed Salamander and Eastern Red-backed Salamander.

Salamander Larvae Comparison

Common Mudpuppy pg 30
Color and pattern: *yellowish dorsolateral stripe.* Toes: *four on hind foot.* Dorsal fin: *only on tail.* Size: *large, reaching a total length of about 6 inches.*

Central Newt pg 32
Color and pattern: *gray or yellow brown with two rows of dorsal light yellow spots; dark stripe through eye.* Toes: *five on hind foot.* Dorsal fin: *nearly to head.* Costal grooves: *none.*

Four-toed Salamander pg 52
Color and pattern: *gray, brown, or greenish with a dark bar behind eye.* Toes: *four on hind foot.* Dorsal fin: *nearly to head.*

Spotted Salamander pg 38
Color and pattern: *brown with dark reticulations or small spots; no blotches on tail fin.* Toes: *rounded, five on rear foot.* Dorsal fin: *nearly to head.*

Tiger Salamander pg 46
Color and pattern: *dark spots on back, possible light stripe.* Toes: *flat and pointed, five on rear foot.* Dorsal fin: *nearly to head.* Size: *large, most reaching a total length around 5 inches.* Cannibalistic form: *larger head with more flattened snout and longer teeth.*

Blue-spotted Salamander pg 36
Color and pattern: *brown with light saddles and crossbars, changing with age to dark blotches on back with black mottling on tail fin; irregular light lateral stripe.* Toes: *rounded, five on rear foot.* Dorsal fin: *nearly to head.*

Common Mudpuppy (subspecies of Mudpuppy)

Necturus maculosus maculosus

Rivers and large lakes. Surprisingly in both turbid slow rivers with mud bottom, and in clear, fast-moving, gravel-bottomed rivers.

Nature Notes:

Mudpuppies swim with an undulating tail motion and legs held at their sides, but usually just walk on the bottom. Lungs are used mostly for depth regulation instead of breathing.

Mudpuppies forage mostly at night, especially in clear water. During the day they hide under rocks or logs to avoid predation.

The Mudpuppy is the only known host for larvae of the Salamander Mussel, *Simpsonaias ambigua*. The tiny larvae, called glochidia, clamp onto the gills of mudpuppies, form cysts, and develop there for 8-12 weeks. Then the cysts break, releasing the tiny mussels. All other known hosts of mussels are fish.

Our largest salamander remains aquatic, breathing with gills all its life.

Description: Length is 13 to 16 inches.

The Common Mudpuppy is our subspecies. It's a ***large salamander with a vertically flattened paddle-like tail***. The skin, which is ***exceedingly slippery and slimy***, varies from brown to gray to nearly black with darker spots or blotches. ***The Mudpuppy has gills, small lidless eyes*** and four toes on each foot.

The gills vary in appearance. Mudpuppies that live in slow moving, stagnant or warm water with less oxygen have large, red, bushy gills. Mudpuppies that live in cool, fast flowing streams with plenty of oxygen have compact, darker gills.

Mudpuppies are neotenic, meaning mature adults are totally aquatic because they retain their larval characteristics, including gills.

The larval stripes are yellow and then turn darker and finally break up and turn to dark brown or black blotches.

Bushy red gills are an adaptation for water low in oxygen. The gills provide much needed surface area for oxygen absorption. Insert: Adults can be nearly black.

Courtship, Mating & Eggs: Reproduction is similar to most salamanders, but Mudpuppies, active all year, breed in fall and early winter. Courtship is followed by deposition of spermatophores which are taken into the female's cloaca. The female lays eggs in late spring or early summer. She hollows out a silt-free nesting area underneath a large flat rock or log. She attaches 18 to 190 eggs, singly, so they hang from the underside of the rock or log (photo below), and then she guards them for about two months until they hatch. The mudpuppy larval stage lasts 4 to 6 years before reaching maturity.

Range Notes: Manitoba south to Tennessee, northeast to Vermont.

Central Newt (subspecies of Eastern Newt)

Notophthalmus viridescens louisianensis

Small ponds, shallow areas of lakes, river backwaters, and clear streams with much aquatic vegetation. Efts in woods near breeding pond.

Nature Notes:

Central Newts are diurnal and forage on various invertebrates plus the eggs and larvae of other amphibians. They are eaten by turtles and frogs. During winter they remain in deeper waters or water kept open by springs, but newts in shallow water bail out and hibernate on land.

Efts are most likely found during or soon after a rain as they walk over leaf litter on the forest floor. Because of its terrestrial habitat, the eft needs more protection and has more poison glands concentrated on its back than the aquatic adult. Some efts are bright orange-red, which is a warning color to potential predators.

The aquatic larva usually becomes a terrestrial eft before developing into an aquatic adult newt.

Description: Length is 2½ to 4 inches.

The Central Newt is a subspecies of the Eastern Newt. It is a two-tone salamander with ***olive-green on the top half, yellow below and speckled all over with black.*** Usually there are no red spots. The skin is dryer than other salamanders and is not slimy. In its normal aquatic adult form, the Eastern Newt has a tail fin but no gills. Adult males, during the breeding season, have an enlarged tail fin that extends from the tail over much of the back. Adult newts forced into a terrestrial existence by environmental conditions lose the tail fin and the skin turns darker and rough. The eft, an intermittent land stage between larva and adult, is reddish brown with rough skin.

The yellow belly is characteristic of both Central Newts and Red-spotted Newts.

The eft is the terrestrial stage of the newt. They are most active during wet weather.

Courtship, Mating & Eggs: Central Newt courtship is in late fall and winter. The male will grasp the female and wave his tail, fanning his scent towards her. Then he walks in front of her and deposits a spermatophore, which she picks up and stores in her cloaca. In April, the eggs are fertilized as the female lays them. She spends several days attaching 12 to 140 eggs, usually singly, on submerged plant stems. Varying with water temperature, the eggs hatch in about a month. By the end of summer the larvae are ready to change. Some metamorphose into the terrestrial eft stage, while others, especially in Minnesota and Wisconsin, metamorphose directly into adults.

Range Notes: Where the Central Newt is found in Minnesota and Wisconsin, its distribution is spotty and the eft stage is rare.

Ranges from our area to the Texas gulf and northern Florida.

salamanders

Red-spotted Newt (subspecies of Eastern Newt)

Notophthalmus viridescens viridescens

Small ponds, shallow lakes, river backwaters and clear streams with much aquatic vegetation. Efts found in woods near breeding pond.

Nature Notes:

Red-spotted Newts are diurnal and forage on various aquatic invertebrates, such as mosquito larvae, plus the eggs and larvae of other amphibians.

During winter they may remain in deeper waters or waters kept open by springs, but newts in shallow water crawl out and hibernate on land.

Because of its terrestrial habitat, the eft needs more protection and has more poison glands concentrated on its back than the aquatic adult. The red or orange coloration of an eft is probably a warning color.

A salamander with little round bright red spots scattered across both the eft and the adult.

Description: Length is 2 ½ to 4 inches

The Red-spotted Newt is a subspecies of the Eastern Newt. It has ***bright red spots, usually bordered by black and sometimes in two rows on the back***. The ***skin is not slimy*** like most salamanders but is dryer. In its normal aquatic adult form, it has a tail fin but no gills. Adult males, during the breeding season, have an enlarged tail fin that extends from the tail over much of the back. If dryer environmental conditions force newts into a terrestrial existence, the tail fin is lost and the skin turns darker and rough. The eft, an intermittent land stage between larva and adult, is red, orange, or reddish brown with a yellowish belly. The red spots are clearly visible on its back

The Red-spotted Newt's eft stage is sometimes bright orange. It is a subspecies of the Eastern Newt found north of Lake Superior in Ontario.

Note the red to orange spots that give this attractive amphibian its name.

Courtship, Mating & Eggs: Active all year with courtship in late fall and winter. The male will grasp the female and wave his tail, fanning his scent towards her. Then he walks in front of her and deposits a spermatophore, which she picks up and stores in her cloaca. In April, the eggs are fertilized as the female lays them. She spends several days attaching 12 to 140 eggs, usually individually, on submerged plant stems. Varying with water temperature, the eggs hatch in about a month. By the end of summer the larvae are ready for a change. Some metamorphose into the terrestrial eft stage, while others metamorphose directly into adults.

Range Notes: Found northeast of Lake Superior in Ontario and also in the southeastern portion of Lower Michigan. Range extends to Nova Scotia, North Carolina and east central Alabama.

salamanders

Blue-spotted Salamander

Ambystoma laterale

Forested areas that have vernal pools for breeding. Tolerant of dry woods and disturbed areas. Even under logs in summer.

Nature Notes:

The Blue-spotted Salamander can exude a white sticky substance from the skin, especially on the base of the tail, to discourage predators. When threatened, the salamander raises and waves its tail, perhaps as a warning.

Blue-spotted Salamanders are more likely to be found under logs during summer than other mole salamanders, most of which have retreated underground.

Our only salamander speckled with blue.

Description: Length is 4 to 5½ inches.

The Blue-spotted Salamander is dark gray to nearly black with ***numerous small blue spots or flecks on the head, body and tail.*** Often, the flecks are heaviest on the sides and tail, with few on the back. Newly metamorphosed salamanders have yellow specks instead of blue.

Courtship, Mating & Eggs: Blue-spotted Salamanders are thought to hibernate around the edge of the breeding pond, because in early spring they often precede the Spotted Salamanders by a couple days. During courtship, the male clasps the female behind her front legs for a time before moving in front of her and depositing one to three spermatophores. She picks up the sperm packet and stores it in her cloaca for a couple days. She lays 1 to 35 eggs, attaching them singly or in

The amount of blue spotting varies. The tail is paler because it was broken off and regenerated.

Older larva show dark blotches on back with mottling on tail fin and an irregular light stripe down each side. Toes are rounded with five on hind foot. Dorsal fin nearly to head.

small clusters to submerged twigs and leaves. The larvae hatch out in 3 to 5 weeks and metamorphose in August.

Range Notes: Absent from extreme western and southwestern Minnesota and a portion of southwestern Wisconsin, but then are found north and east through Michigan. Range extends from eastern Manitoba to Nova Scotia and to northeast Illinois.

Eggs attached to submerged twig.

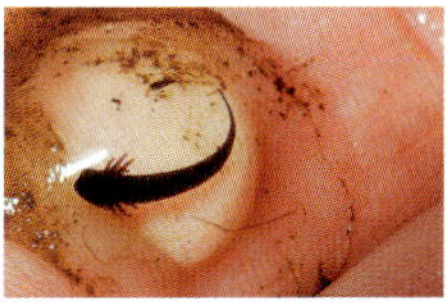

Developing embryo inside egg.

Still a juvenile, but its original yellow flecks have now turned blue.

Spotted Salamander

Ambystoma maculatum

Closed canopy deciduous or mixed forests. Vernal ponds are vital for reproduction.

Nature Notes:

Spotted Salamanders are rarely seen except during breeding and migration times. They spend most of the time burrowed under ground, where they keep cool, moist, and well fed on worms and other invertebrates.

Must have closed canopy forests to retain needed moisture on the forest floor.

A male deposited these spermatophores on a twig.

This salamander has rounded yellow spots, sometimes with a pair of orange spots on the back of its head.

Description: Length is 5 to 7 inches.

The Spotted Salamander is ***dark gray to nearly black with yellow spots loosely arranged in two rows on the body and scattered on the tail, legs and head.*** Some individuals have only a few spots or none at all. Often there is an orange spot behind each eye. Newly metamorphosed Spotted Salamanders are brown with tiny gold or yellow flecks.

Courtship, Mating & Eggs: In late March or early April, rain triggers the migration of Spotted Salamanders to their breeding ponds. Males arrive a day or two before females. Courtship precedes the deposition of spermatophores, which the female retrieves and stores in her cloaca. A day or two later she

This recently metamorphosed salamander will develop its yellow spots.

Adults sometimes have a pair of orange spots (sometimes four) on the head.

The Spotted's firm egg mass.

lays her 50 to over 250 eggs in one or more masses attached to submerged sticks or stems. After some swelling with water, the result is a tennis ball sized mass of firm clear or cloudy substance with eggs embedded inside. It holds its shape if removed from water. It takes the eggs 20 to 60 days to hatch, depending on water temperature. The larvae take at least two months to develop, with the option of over-wintering if the pond doesn't dry.

Range Notes: Not known in Minnesota until 2001, when its egg masses were located in Pine County in the Nemadji State Forest.

Range is Ontario to Quebec and south to South Carolina and west to east Texas.

Gray Tiger Salamander (subspecies of Western Tiger Salamander) *Ambystoma mavortium diaboli*

Prairies, fields and woodlands near wetlands, small lakes and ponds.

A salamander that sometimes retains its gills and remains aquatic into adulthood.

Description: Length is 6 to 8½ inches.

The Gray Tiger Salamander is the subspecies in our area. It varies in its coloration and markings. It can be black, gray, brown or olive colored with markings that are black, brown, green or yellow. The size of the markings also vary from small spots to large bold marks and are found across the back and on head, tail, and legs. The belly is lighter in color with fewer markings. On the bottom of their feet are hardened tubercles which are used for digging into the ground.

The larvae have tail fins reaching almost to their neck and show great variability in the amount and distribution of heavy black markings on their grayish-green bodies.

Nature Notes:

Gray Tiger Salamanders frequently develop into adults while retaining their gills and other aquatic characteristics. Since these gilled adults can't very well hike up into the woods and burrow in for the winter, they survive in waters that are deep enough to not freeze. Those that develop normally, leave the pond like most other salamanders and burrow below the frost line to spend the winter.

Gray Tiger Salamanders sometimes live over 25 years.

Coloration is quite variable. Compare with the Eastern species of Tiger Salamander.

Courtship, Mating & Eggs: Early spring thundershowers spur the salamanders from their winter burrows and they move to fish free breeding sites in shallow lakes, wetlands or flooded fields. After a brief courtship, the female deposits eggs in small gelatinous masses attached to submerged sticks or plant stems. The total number of eggs can range from a hundred to 5,000. They hatch in three to five weeks. The half inch larvae that hatch have no limbs but develop them by the time they double their size. Most larvae metamorphose into adult salamanders by fall, but some retain their gills and remain aquatic.

Diet: Earthworms, insects and spiders.

Range Notes: In our region only in west central Minnesota.

From Minnesota, found west and northwest into southwest Manitoba and Saskatchewan.

salamanders

Marbled Salamander

Ambystoma opacum

Floodplain forests, wooded areas, wooded or rocky hillsides.

Nature Notes:

Like other mole salamanders, the Marbled Salamander is secretive and rarely seen. But unlike the others, it is completely terrestrial and doesn't enter ponds. Its courtship and egg laying are in the fall and on land. By spring, these larvae have grown and developed enough to have an advantage over the larvae of spring breeders. The larval Marbled Salamanders can out compete other larvae and even feed on them.

The Marbled Salamander has tiny glands in the skin that are concentrated in the top of the tail, and they secrete noxious substances. They sometimes raise the tail at a would-be predator, which may also deflect an attack to a less vital area.

Our only salamander that breeds on land and deposits eggs where autumn rains will submerge them.

Description: Length is 3½ to 5 inches

The Marbled Salamander is a ***stocky salamander that looks well-fed.*** It is black or dark brownish ***black with white or gray bands across the back, tail, and head.*** Often the bands end at a matching white or gray side stripe. Males have the white banding and females have gray banding.

The larvae are dark brown with light markings on the sides and tail fin. Newly metamorphosed salamanders have yellow markings which turn silver-gray or white when they emerge from the water. It takes several weeks to develop the unique adult pattern.

Courtship, Mating & Eggs: Breeds in the fall, which is unusual among salamanders.

Completely terrestrial, the Marbled Salamander is not found in ponds like other salamanders.

Does the Marbled Salamander still exist in southwest Michigan? Hopefully field researchers will locate undiscovered populations.

Adults migrate to the edge of dried or drying vernal ponds or other depressions and hide under logs, bark or leaves. The male deposits a spermatophore and the female picks it up in her cloaca. She then lays 50 to 200 eggs underneath a rotting log or piece of bark or moss. She guards her eggs until rains flood them. Then they hatch and finish their development submerged in the rain-filled pond.

Range Notes: Recorded in our area only from three counties in southwest Michigan where it was last found in Allegan County in 1989. It is listed as endangered but is probably extirpated.

Range extends south to east Texas and barely into Florida and east to New Hampshire.

Small-mouthed Salamander

Ambystoma texanum

Floodplain forests with vernal ponds. Also farmland and prairies with shallow ponds, river backwaters and flooded fields for breeding.

Nature Notes:

Glands on the upper tail secrete a toxic or distasteful substance and the salamander will raise its tail and may even whip it towards a predator. If that doesn't work, some Small-mouthed Salamanders will try a stiff defensive posture with head and tail raised high and their legs held tight against their sides. This behavior, used by some other amphibians as well, is called the Unken reflex. Some predators seem to lose interest in the now stiff and still salamander.

Small-mouthed Salamanders will frequently hybridize with Blue-spotted Salamanders and other mole salamanders.

Besides its small mouth, this salamander has a small head and a slightly stuck out lower jaw.

Description: Length is 4½ to 7 inches.

The Small-mouthed Salamander has a ***small head and a short snout.*** It's dark colored with light scattered markings, especially on the sides. In a profile of the head, notice that ***the lower jaw is stuck out slightly farther than the upper jaw.***

Courtship, Mating & Eggs: Early spring rains send the adults to breeding ponds where a bumping and nudging courtship results in males depositing spermatophores on submerged twigs or the pond bottom or on top of another male's spermatophore. The female takes up one or more spermatophores into her cloaca and attaches small masses of eggs to submerged plants but totaling 300 to 700 eggs. The eggs hatch in two to eight weeks, depend-

Look close and notice the lower jaw sticks out farther than the upper jaw.

ing on water temperature, and then the larvae need two to three months of development before metamorphosing into a terrestrial adult.

Range Notes: In our three states, the Small-mouthed Salamander is rare and found only in southeastern Michigan. Unfortunately, that area has lots of people and pressure to develop on or around the few breeding ponds.

From extreme southeast Michigan its range extends into Ohio and west through Indiana and Illinois and across southern Iowa to eastern Texas and to western Alabama.

Nature Notes:

Except when breeding from late February through March, Small-mouthed Salamanders retreat into crayfish burrows or mammal burrows or underneath rotting logs and stay out of sight although, they may venture out on rainy nights.

Eastern Tiger Salamander (subspecies of Tiger Salamander) *Ambystoma tigrinum tigrinum*

Woods, open areas and even suburbs. Breeding areas include forest ponds, farm ponds, gravel pits and fish rearing ponds.

Nature Notes:

There are a few populations of Tiger Salamanders in our area that are neotenic. These salamanders reach sexual maturity without metamorphosing into the adult form. They retain their gills and tail fins and remain aquatic.

Since Tiger Salamanders live underground and migrate on rainy nights, they often escape notice until they fall into someone's window well or swimming pool.

This salamander is less particular than other *Ambystoma* about breeding ponds, using woodland ponds, farm ponds, gravel pits, and even fish-rearing ponds.

Some larvae of this salamander develop enlarged heads and turn cannibalistic.

Description: Length is 7 to 8 inches with a record of 13 ¾ (Minnesota specimen).

The Eastern Tiger Salamander is the subspecies found in our area. It is ***dark gray or black with highly variable and random yellow blotches, spots and bars that extend to the belly***. The ***lower lip and throat are yellow***. Newly metamorphosed salamanders are spotted with yellow and gradually add more yellow markings.

Courtship, Mating & Eggs: Tiger Salamanders migrate to breeding ponds in early spring, with males arriving two or three days before the females. Courtship involves a male nudging at a female, followed by his deposition of a spermatophore and the female taking it into her cloaca. After about a day the female will lay her fertilized eggs in clusters

Eastern Tiger Salamanders are always dark gray or black but they are highly variable in the amount of yellow spotting. Inset: Some individuals are marked with large yellow blotches.

attached to twigs, leaves, or other debris near the bottom of the pond. They don't have the same firmness as those of spotted salamanders and fall apart when lifted. The eggs hatch in two to four weeks. Larval development is affected by water temperature, water quality, larval density, and the permanence of the pond. In a permanent pond, the larvae can overwinter. In a drying pond, the density of salamander larvae can greatly increase. Larvae respond by speeding up their rate of development.

Migrating to the breeding pond on a rainy night.

The shed skin of a molting Tiger Salamander.

Evolution came up with an additional solution in some Tiger Salamander populations. Some larvae become cannibals. They were already carnivores, but now they develop larger heads, bigger mouths, longer teeth and proceed to devour their siblings.

Range Notes: Extreme northwest Minnesota to Kentucky and scattered populations beyond into east Texas, Florida, South Carolina and New York.

salamanders

Northern Dusky Salamander

Desmognathus fuscus

Edges of rocky streams and springs in wooded areas. Hide under flat rocks adjacent to or partly submerged in water.

Nature Notes:

Those large back legs are used for jumping. I learned that when I pulled the cover off a container with several inside. Suddenly they were jumping out like popcorn! To add to the difficulty, they are also very fast and slippery.

The Northern Dusky Salamander hides under flat rocks adjacent to or partly submerged in water.

The Plethodontidae, the family of lungless salamanders including the Northern Dusky Salamander, is unique in having an immovable lower jaw and it must lift its head to open its mouth.

The Michigan location is a clear stream in an Eastern Hemlock forest.

This salamander escapes predators by jumping and can lose its tail followed by regenerating a new one.

Description: Length is 2½ to 4½ inches.

The Northern Dusky Salamander is best identified by the ***light stripe running from the eye to the rear of the mouth*** and by the ***rear legs being larger than the front legs.*** Coloration is variable but it often has a broad stripe of tan, light brown or yellowish brown on its back running from head to tail. This broad stripe is flanked by dark, wavy markings. Old adults may be all dark.

The tail is vertically flattened near the tip with a sharp top edge.

The larva has 5 to 8 pairs of spots that can be tan, yellow or reddish orange.

Often near water, this salamander can escape danger by using speed and jumping ability to escape into the stream.

Coloration is variable but look for the light stripe running down the back.

Courtship, Mating & Eggs: Breeding is during the summer with courtship taking place on land. The female picks up the spermatophore sperm packet and deposits her 12 to 27 eggs inside a rotted log or in a moist hollow near a stream or on the underside of submerged rocks. She guards the eggs for roughly two months before they hatch and the young move to the stream to complete their development

Range Notes: The Northern Dusky Salamander has been found in our area only from one isolated population in eastern Lower Michigan, and it is rare.

Ranges from southeast Indiana to halfway into South Carolina and northeast to Maine and Quebec.

Southern Two-lined Salamander

Eurycea cirrigera

Rocky streams, springs, and seepages with continuous flowing water in hardwood forests.

Nature Notes:

Southern Two-lined Salamanders may stay active all year if they have running water, and they need permanent running water for their larvae. They are a streamside species, hiding under flat rocks beside a stream, but they also wander into the woods away from water.

They are fast and often escape predators with their rapidly undulating body. They can jump by curving the body and suddenly straightening out. If grabbed, they can break off part of their tail, which wiggles and may distract a predator long enough for the salamander to evade capture. The missing end of the tail will be regenerated.

Breeding males of this salamander have a slender cirrus, or extension, hanging down from each nostril.

Description: Length is 2½ to 3¾ inches.

The Southern Two-lined Salamander is slender and light ***yellowish brown or bronze with black dorso-lateral stripes running down the sides of the back.*** In between the stripes is a sprinkling of black spots or flecks. Males have a slender extension, called a cirrus, extending down from each nostril to just below the mouth.

The larvae are dark with two rows of light-colored spots on the back.

Courtship, Mating & Eggs: Southern Two-lined Salamanders breed in late March or April. The male nudges and sniffs at the female and deposits a spermatophore. The female picks it up with her cloaca. Then she attaches a single layer of 20 to 100 eggs to the

Note the black stripes running down the upper sides of the back. In the bottom photo notice the regenerated end of the tail.

underside of flat rocks, usually in running water. She attends the eggs until they hatch in May or June, and then the larvae complete their development in the water, which takes one or two more years.

Nature Notes:

The purpose of the male's cirri under the nostrils is probably to sense a mate or a male rival or to help identify food.

Range Notes: In our area, it is only found in one isolated area of eastern Lower Michigan. Since this is a relatively recent discovery, it isn't known yet if they have been here all along or if they were introduced.

Main range is Indiana south to Mississippi and east and north into Virginia.

salamanders

Four-toed Salamander

Hemidactylium scutatum

Forests with wet depressions and sphagnum moss. Spring-fed swamps, spring-fed bogs and spring-fed creeks.

Nature Notes:

The Four-toed Salamander can break off its tail as a defense. The wiggling tail may distract a predator long enough for the salamander to escape. A new tail is regenerated.

Very secretive and difficult to find. Their habitat requirements severely limit where they can survive, and even in optimum habitats they are never plentiful.

This salamander's tail is constricted at the base and will break off if grabbed by a predator.

Description: Length is 2 to 3½ inches

The Four-toed Salamander is named for having ***four toes on each hind foot***, since most salamanders have five. This salamander has a ***reddish-brown back, often spotted with black.*** Its gray or brown sides are speckled. The base of the tail is constricted, and the ***belly is white with black dots.*** Juveniles have darker bellies with blue dots. Newly metamorphosed Four-toed Salamanders have a short dark stripe behind each eye.

Courtship, Mating & Eggs: Courtship takes place during fall or spring. The male rubs his nose and chin against the female's snout. The female straddles the male's tail and nudges the base of his tail. The male deposits a spermatophore, and the sperm packet is picked up and stored in the female's cloaca. During

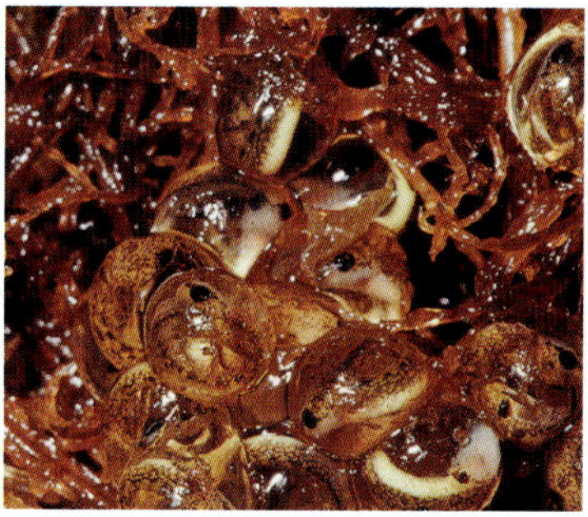

Eggs are laid in sphagnum moss that overhangs the water or mounds at the base of willows and alder in or adjacent to the water.

Note the black spots on the belly. Tail is constricted at base and will break off.

Sphagnum moss is an essential habitat requirement.

spring, the female migrates to a site for her eggs. The ideal situation has sphagnum moss overhanging shallow, spring-fed waters. If not actually overhanging the water, there must be ground vegetation or a rotted log next to the water. The female, in a small shelter or hollow, turns upside down and lays a cluster of 15 to 64 eggs. The eggs stick to each other and to the ceiling of the hollow. The female guards them for 38 to 60 days until the eggs hatch. The hatchling larvae drop into the water or wriggle until gravity helps them get there. They need three to eight weeks to transform into the terrestrial form.

Range Notes: This species was first discovered in Minnesota in 1994 in Itasca County. In 1999, more were found in Aitkin, Carlton, Mille Lacs and Pine Counties. St. Louis County has since been added to the list. These discoveries were made by Carol Hall and her Minnesota County Biological Survey teams.

Ranges east to Maine and southwest to Alabama and scattered into Arkansas.

Eastern Red-backed Salamander

Plethodon cinereus

Forests with closed canopy and leaf litter, rotting logs, or moss. Terrestrial. Does not need standing water for breeding.

Nature Notes:

Secretive and hide during the day. On damp nights they emerge to search for invertebrates to eat, often climbing on the trunks of trees or other vegetation.

Eastern Red-backed Salamanders are territorial. Both male and female will defend their territory, probably as a defense of their food source.

The salamander raises its body off the ground to threaten an intruder. The threat is increased when it raises its tail also. If necessary, it bites at the intruding red-back's nasolabial groove or tail to drive it away.

A terrestrial salamander that lays eggs on land and has no aquatic larval stage.

Description: Length is 2 ½ to 4 inches

The Eastern Red-backed Salamander is small and has small legs with five toes on each hind foot. ***Most of our specimens are the "redback" color phase, with a brick red color straight down the back and onto the tail.*** The sides and belly are gray with light and dark speckling. There is also a "leadback" phase, which is very dark all over with possible light sprinkling.

Courtship, Mating & Eggs: During fall or spring, the male and female engage in courtship with the male rubbing his chin over the female's snout and the female straddling the male's tail and nudging the tail base. The male deposits a spermatophore, which the female takes into her cloaca. In June the female finds a small hollowed, moist place, such as inside or

Female guarding her eggs (left). Eggs are often laid inside a rotting log.

underneath a rotted log. She lays 3 to 17 eggs which are often hung from above by a thin gelatinous stalk. The female guards the eggs for one to two months until they hatch. The little hatchlings have relatively big heads, plus tiny remnant gills that are soon absorbed; otherwise they look like the adults.

Searching for prey on a tree trunk at night.

Range Notes: The leadback phase is very rare or does not occur in Minnesota or Wisconsin. In Michigan, the leadback phase can be found on the Upper Peninsula only in the area of the Straits of Mackinac. In Lower Michigan, the leadback phase is found in about 10 percent of the population.

U.S. range from Minnesota east to Maine and southwest to Indiana and North Carolina.

salamanders

Western Lesser Siren (subspecies of Lesser Siren)

Siren intermedia nettingi

Ponds, swamps and ditches with muddy bottoms and lots of aquatic vegetation.

Nature Notes:

During droughts they burrow into the mud or retreat into crayfish burrows. They create a mucous cocoon for themselves that covers all but the mouth to conserve water as the mud dries. Their metabolism drops and adults can aestivate for over a year, if necessary.

The Western Lesser Siren was first discovered in Michigan in 1958 when several dead individuals littered the shore of a lake treated with Rotenone. They were listed as a Species of Special Concern during the 61 years that no one found any more. In 2019 and 2020 they were found again in Michigan in Van Buren County. They are nocturnal and extremely secretive and spend the day buried in the mud bottom.

This unique salamander feeds by sucking in water and prey like an aquatic vacuum cleaner.

Description: Length is 7 to 19 ¾ inches.

The subspecies found in our area is the Western Lesser Siren. It's an ***aquatic, slimy, eel-like salamander with no hind limbs***, but it has tiny front legs with four digits on each. The body is usually brown to gray in color with small, scattered black spots, but it may be all black.

It has external red gills. The male has a blunt snout and head with large jaw muscles. The female has a longer, flatter snout.

Young larvae have red or yellow markings on the head and a dorsal fin which gradually disappears between month two and nine.

The Western Lesser Siren may travel overland on wet nights.

Courtship, Mating & Eggs: Fertilization is external and 12 to 300 eggs are laid in depressions in the bottom mud of calm water. Upon hatching, the larvae are less than a half inch long, and it takes them two to three years to reach maturity.

Range Notes: In our area, the Western Lesser Siren is known from only two counties in southwestern Lower Michigan.

Main range is Indiana south to Mississippi and east and north into Virginia.

Nature Notes:

Sirens communicate by rapidly snapping their jaws shut and producing a clicking sound. They can also make other sounds like squeaks and whistles and will yelp if grabbed by a predator.

Western Lesser Sirens can move overland at times and colonize ponds and other unconnected waters.

Order Anura: Frogs & Toads

Adult frogs and toads have a big head with no distinct neck, a short body with no tail, and long muscular hind legs. Frogs and toads have big eyes with eyelids, and a tympanum (for hearing) behind each eye. They have lungs for breathing, but oxygen can also be absorbed through the skin. Frogs and toads have long sticky tongues to help capture prey.

There are about 4,200 species of frogs and toads in the world, which are divided into about 26 families. Sixteen species in three families are found in our three northern states.

Family Bufonidae – True Toads

Over 400 species of toads in the world are in the family Bufonidae. Toads have thick, dry, warty skin that resists dehydration and produces irritating toxins. The glands producing the milky white poison are distributed across the toad's back, but are concentrated in the two parotoid glands on the back of the head. Toads have relatively short hind legs which are used for walking, hopping, or burrowing by backing into the ground. Four species of toads inhabit our area: the American Toad, Great Plains Toad, Fowler's Toad, and Canadian Toad.

Family Hylidae – Treefrogs

The family Hylidae has about 720 species worldwide. Treefrogs have toes with expanded adhesive discs called toe pads, which allows an arboreal life style. There are six species in our area: Blanchard's Cricket Frog, Cope's Gray Treefrog, Gray Treefrog, Spring Peeper, Boreal Chorus Frog and Western Chorus Frog.

Family Ranidae – True Frogs

About 627 species of frogs in the world are grouped in the family Ranidae. These are the typical jumping frogs with long legs and webbed feet. They have large tympanums, and most have dorsolateral ridges, which can be useful for identification. Our three state area has six species of these frogs: American Bullfrog, Green Frog, Pickerel Frog, Northern Leopard Frog, Wood Frog, and Mink Frog.

Toad cranial crests & parotoid glands

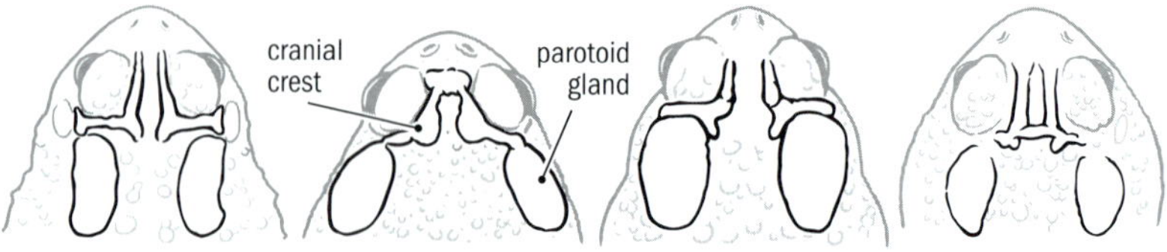

American Toad (pg. 60) **Great Plains Toad** (pg. 62) **Fowler's Toad** (pg. 64) **Canadian Toad** (pg. 66)

Tadpole Comparison

American Toad pg 60
Color and pattern: *black or dark brown, often with bright metallic spots.* Tail fin: *clear, rounded at tip.* Tail musculature: *no pigment on bottom half.*

Gray Treefrog pg 72
Tail fin: *high, extending onto upper body, red tinge, mottled with black, ends in flagellum.* Tail musculature: *evenly pigmented.*

Spring Peeper pg 74
Color and pattern: *gray, brown, or greenish with a dark bar behind eye.* Toes: *four on hind foot.* Dorsal fin: *nearly to head.*

Chorus Frogs pg 76, 78
Color and pattern: *brown with dark reticulations or small spots; no blotches on tail fin.* Toes: *rounded, five on rear foot.* Dorsal fin: *nearly to head.*

American Bullfrog pg 80
Color and pattern: *dark spots on back, possible light stripe.* Toes: *flat and pointed, five on rear foot.* Dorsal fin: *nearly to head. Size: large, most reaching a total length around five inches. Cannibalistic form: larger head with more flattened snout and longer teeth.*

Green Frog pg 82
Color and pattern: *brown with light saddles and crossbars, changing with age to dark blotches on back with black mottling on tail fin; irregular light lateral stripe.* Toes: *rounded, five on rear foot.* Dorsal fin: *nearly to head.*

Leopard Frog pg 86
Color: *green, olive, or brown.* Tail fin: *light or clear, weakly spotted with black.*

Wood Frog pg 90
Color and pattern: *greenish brown, often speckled with black and gold, belly iridescent whitish to bronze.* Tail fin: *short and high, light with dark spots.*

Eastern American Toad (subspecies of American Toad) *Anaxyrus americanus americanus*

Forests and open woodlands with moist soil for burrowing. Shallow ponds for breeding (flooded ditches to huge marshes).

Nature Notes:

Because this toad is terrestrial, it needs plenty of defenses. It has excellent camouflage and texture, burrows into the ground or hides under logs and when threatened, it lowers its head, presenting its toxic parotoid glands to the predator. Further stressed, the toad secretes a milky, white, toxic substance from skin glands. When seized by a snake that swallows prey whole, the toad inflates its body larger than the snake can swallow.

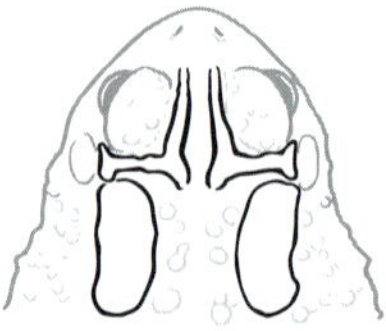

Dorsal view of head showing cranial crests & parotoid glands.

A toad that vocalizes with long (up to 30 seconds) steady musical trills.

Description: Length is 2 to 3 ½ inches.

The Eastern American Toad is the subspecies in our area. It has the following features: *cranial crests (ridges on top of head) that do not touch the parotoid glands or only the tip touches, the largest dark spots contain only one or two large warts, the tibia (leg between thigh and foot) has enlarged warts and the chest is spotted.* It can be gray, dark gray, brown, reddish brown or more, but will likely resemble an environmental constant like soil color.

Voice: A steady trill lasting about 30 seconds.

Courtship, Mating & Eggs: During late April or early May, American Toads migrate to shallow, temporary breeding ponds. Males arrive first, by as much as two weeks. The male sits upright in shallow water or even on the

Male's mating call is a steady trill lasting up to 30 seconds.

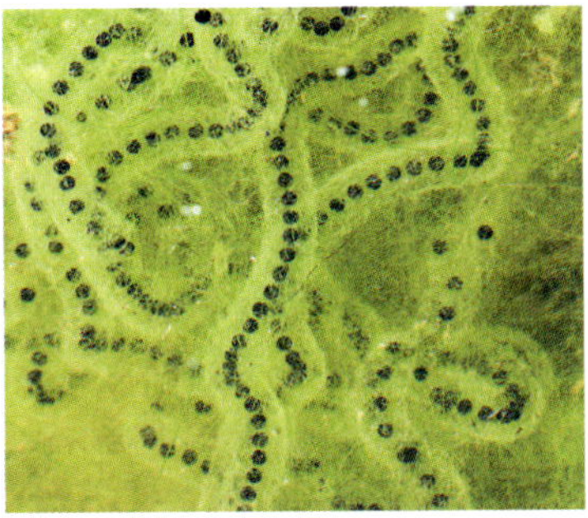

Thousands of eggs are laid in long pairs of gelatinous strings.

This toad is hiding in a shallow burrow.

Toads feed on worms, insects, spiders, slugs.

shore, so his large vocal pouch will be out of the water. His call is a steady trill lasting about 30 seconds. Any other toad that happens by will be grabbed. If a male clasps another male by mistake, the latter gives a one to five second release call trill while vibrating his sides. When he amplexes a female, enlarged thumbs help him maintain a tight grip if another male tries to dislodge him. The female swims to an egg laying site with the male hanging on her back. The male fertilizes 2,000 to over 20,000 eggs as the female deposits them in a pair of long gelatinous strings. Small black tadpoles hatch within two weeks and metamorphose in 6 to 10 weeks. In mid-summer, the mud flats around a drying pond may be hopping with hundreds of toadlets.

Tiny toadlets look similar to the adult.

Range: Canada, eastern Dakotas and eastern. Kansas to North Carolina and north Georgia.

toads

Great Plains Toad

Anaxyrus cognatus

Open areas like prairies, grasslands, fields, and floodplains.

Nature Notes:

The Great Plains Toad's toxic skin is a defense against most predators but not against the Plains Hog-nosed Snake. When seized by the snake, the toad puffs up his body with air so he is too large to be eaten. But the snake has a pair of enlarged rear teeth that puncture the toad and it is swallowed whole.

Dorsal view of head showing cranial crests & parotoid glands.

This toad's call is up to 50 seconds, but it's pulsating and mechanical and not musical.

Description: Length is 2 to 3½ inches.

The Great Plains Toad has ***dark spots on its back with light colored edging*** around each one. ***Each spot has several, small reddish warts.*** The ground color is tan to dull green. The ***parotoid glands are oval and smaller than those of American Toads.*** The ***cranial crest is V-shaped between the eyes.*** The ***belly is white*** with no black markings, but adult males have a dark throat. The ***male's vocal pouch is oblong when calling*** and protrudes farther than the American Toad's rounded vocal pouch.

Voice: A harsh, loud, mechanical call similar to a jackhammer for nearly a minute.

Courtship, Mating & Eggs: Great Plains Toads depend on early summer thundershowers to provide water for breeding. Then they

This Great Plains Toad is obviously a female because her throat is white instead of dark.

quickly emerge from the ground and head for the new shallow breeding pools. A female can lay over 40,000 eggs and breed more than once in a summer. In only two to seven days the young are hatched. Their tadpole stage is temperature dependent and varies from nearly three to six weeks.

Range Notes: Western Minnesota into Montana and south to west Texas, Arizona and Mexico.

Nature Notes:

The Great Plains Toad is mostly nocturnal and an expert burrower, both of which help it avoid summer heat. But when the weather remains hot and dry the toads burrow down and estivate until it rains.

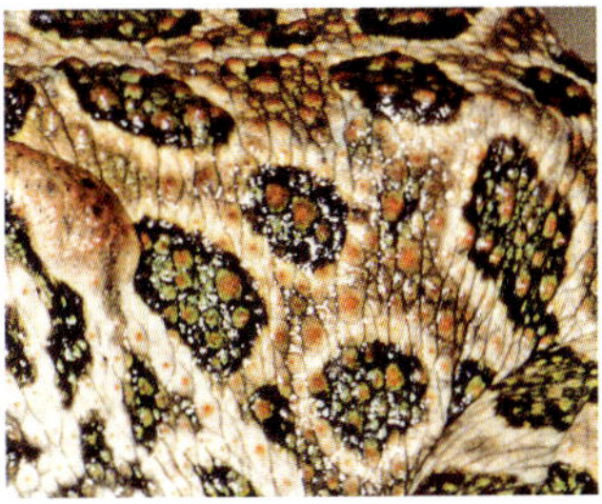

Each spot encircles several reddish warts.

Fowler's Toad

Anaxyrus fowleri

Open woodlands and open areas (avoids dense forests). Needs moist soil for burrowing and leaf litter and debris for hiding.

Nature Notes:

Fowler's Toad has all the defenses of the American Toad, but it might have one more. Reportedly it will sometimes play dead when roughly handled. Some predators will not feed on prey that is already dead.

Fowler's Toad is more active than the American Toad during daylight hours but will still burrow into the ground and estivate during hot, dry weather.

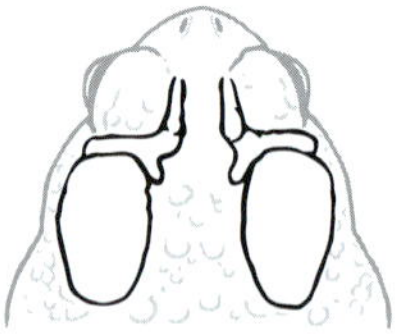

Dorsal view of head showing cranial crests & parotoid glands.

A toad that calls with a low bleat only lasting two to seven seconds.

Description: Length is 2 to 3 inches.

Fowler's Toad can be distinguished from the American Toad in several ways. First, consider range: Fowler's Toad does not occur in Minnesota, Wisconsin or Michigan's Upper Peninsula. The ***parotoid glands of Fowler's Toad are up against the cranial crests, not just touching a tip***. The ***largest black spots on the back contain three or more warts***. The tibia does not have greatly enlarged warts. The chest has no spots or perhaps one.

Voice: The male Fowler's Toad advertisement call, a low pitched nasal "*waaah*," bears no resemblance to the American Toad's call.

Courtship, Mating & Eggs: Fowler's Toad uses a variety of temporary pools for breeding, which usually occurs later than the breeding

The Fowler's Toad's color doesn't match this background because it was crossing a gravel road.

Note that the largest black spots on a Fowler's Toad have three or more warts. American Toads only have one or two warts within a black spot.

of American Toads in the same area. The female Fowler's Toad averages 7,000 to 10,000 eggs, deposited in two long strings. Inspection with a magnifying glass shows the egg-holding strings have no partitions, distinguishing them from the strings of American Toads where partitions keep eggs separate. The tiny black tadpoles, which look like American Toad tadpoles, hatch out in a few days and metamorphose about a month later.

Range Notes: From western Michigan, Fowler's Toad ranges to Louisiana and east into North Carolina and north New England.

toads

Canadian Toad

Anaxyrus hemiophrys

Woodlands and wetlands usually near a body of water.

Nature Notes:

The Canadian Toad is more aquatic than other toads and will swim out into the water to escape danger.

Most hibernate communally in mounds called Mima mounds. The mounds are believed to be formed by gophers and are composed of soil that is looser, allowing easier digging for toads. As winter progresses, the toads burrow to keep below the frost line. A few toads hibernate individually in wooded uplands.

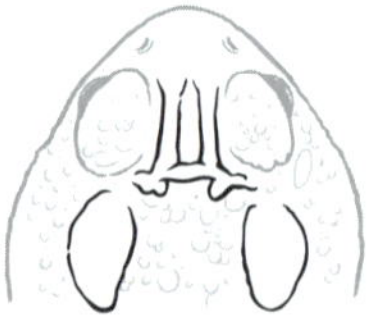

Dorsal view of head showing cranial crests & parotoid glands.

These toads hibernate together with hundreds in the same small earthen mound.

Description: Length is 2 to 3 inches.

The Canadian Toad looks like an ordinary American Toad at first glance, but can be easily distinguished. First consider where it was found. Canadian Toads only enter our area in northwestern Minnesota. Second, look at the top of its head. The Canadian Toad has a ***large boss or bump between its eyes, formed by the merging of the cranial crests.*** The parotoid glands of the Canadian Toad are not well defined, blending into the skin around the edges.

Voice: The advertisement call of the Canadian Toad is a low-pitched, weak trill lasting only two to five seconds.

Courtship, Mating & Eggs: Breeding takes place in in the shallows of ponds or lakes and

Canadian Toads can be distinguished from American Toads by the large bump between its eyes.

the female lays strings of 3,000 to over 5,000 eggs which hatch a few days later. The toadlets emerge from the water in late June or early July.

Range Notes: The Canadian Toad is found in only northwest and west-central Minnesota in our area.

From northwest Minnesota its range extends into the eastern Dakotas and northwest into Manitoba, Saskatchewan and Alberta.

Blanchard's Cricket Frog

Acris blanchardi

frogs

Edges of streams, rivers, lakes and ponds; especially those with vegetation and mud flats or areas of gravel.

Nature Notes:

For those of you used to watching little toads hopping, cricket frogs will amaze you with their long leaps. Their back legs are strong and used for escaping. If they leap into the water, they avoid deep water and swim a half circle back towards the shore.

Adults become inactive during August, but the young of the year are still growing and remain active into September.

A tiny frog with warty skin and long-leaping escapes.

Description: Length is 5/8 to 1½ inches.

Blanchard's Cricket Frogs have ***numerous, small wart-like bumps*** on their brown, green or gray skin. Often there is a ***dark triangle between the eyes pointing to the rear***. Some have a ***reddish brown or green blotch on the middle of their back*** and sometimes on the snout as well. Although they are treefrogs, the toepads are hardly noticeable.

The tadpoles are the only ones in our area with black tail tips.

Voice: The male's call is a *"gik gik gik"* ...like two pebbles hitting together.

Courtship, Mating & Eggs: Breeding is in late May, June and July in permanent bodies of water. The female attaches 15 to 20 eggs at a time to submerged vegetation and may lay

Back markings aren't always green.

Vocalizing in late spring/early summer.

Blanchard's Cricket Frogs demonstrating their various colors.

up to 400 eggs in a season. They hatch a few days later, and it takes the tadpoles five to ten weeks to metamorphose into adults.

Range Notes: Years ago they were in the southern thirds of both Wisconsin and Michigan but populations sharply declined or disappeared for unknown reasons. Lately they have been making a slow comeback.

Southern Michigan and Ohio west to Wisconsin and southeastern Minnesota and southwest into Texas.

Cope's Gray Treefrog

Hyla chrysoscelis

Woodland edges and more open country. They avoid deep forests.

Nature Notes:

Treefrogs are often seen feeding on insects attracted by lit windows and porch lights. During the day, treefrogs hide in whatever nearby cavities are available, or they employ their protective camouflage and just hunker down on a leaf or against a tree trunk.

Apparently Cope's Gray Treefrog never climbs more than ten feet high in a tree, because they are never heard calling from a higher perch.

This treefrog turns bright green when vocalizing with its short harsh trill.

Description: Length is 1¼ to 1¾ inches.

Cope's Gray Treefrog has slightly warty skin, slender legs and enlarged toe pads. Often a white, roughly square spot is under the eye, and a dark band extends from the eye to the front leg. Cope's Gray Treefrog can change color to gray or green or a mix in between. The groin and undersides of the rear legs are often washed with yellow or orange. Unfortunately, this description also fits the Gray Treefrog, *Hyla versicolor*. In fact these two species were thought to be the same species until 1968. They are distinguished by their advertisement calls, which are similar, but different. ***This species call is faster and shriller than the musical trill of H. versicolor.*** They are also differentiated by chromosome number (*H. versicolor* is tetraploid, twice the normal num-

Best separated from Eastern Gray Treefrog by call. Cope's Gray Treefrog changes color for camouflage purposes but is often all green without blotching during the breeding season. Juveniles of both gray treefrog species are bright green.

ber), but that's not helpful for identification in the field. There are some visible differences that have emerged. *Cope's Gray Treefrog has smoother skin. Often there is no pattern and the male tends to be all green when calling. If there is a pattern, Cope's Gray Treefrog has no black border around the blotches.* The average length of Cope's Gray Treefrog is 1 ½ inches, while the average Gray Treefrog is 1 ¾ inches. Recently metamorphosed treefrogs of both species are bright green.

Voice: The call is a short, harsh trill.

Courtship, Mating & Eggs: About the middle of May, Cope's Gray Treefrogs begin calling from various small, ponds or the shallow edges of lakes. The males prefer perches just over the water and will defend territories. Satellite males lurk nearby to replace the resident male if he finds a female and leaves his perch. Females lay 1,000 to 2000 eggs in small clusters attached near the surface. The eggs are hatched in a week, and the tadpoles metamorphose six to eight weeks later.

Range Notes: From Minnesota south into Texas and east into the Florida panhandle and north to New Jersey.

Range is uncertain in some areas, since it was originally combined with that of the Gray Treefrog. Ranges sometimes overlap. Both species sometimes call from the same ponds.

frogs

Gray Treefrog
Hyla versicolor

Deciduous, coniferous, and mixed forests. Unlike Cope's, often found in deep forests.

Nature Notes:

Sometimes a Gray Treefrog can be heard calling from high in a tree, before or well after breeding season. Some people mistake it for a bird. This is *Hyla versicolor*, because Cope's Gray Treefrog apparently doesn't call from over ten feet high.

Gray Treefrogs can change color from green to gray to blend in with their substrate.

A treefrog that adds a beautiful, musical trill to a warm May night.

Description: Length is 1½ to 2 inches.

The Gray Treefrog can be gray or green, often with blotches. Usually adults have ***a roughly square white spot under the eye*** and a wash of yellow on the concealed surfaces of the back legs. Of course it looks very similar to Cope's Gray Treefrog, *Hyla chrysoscelis*. The two are best distinguished by their calls. The Gray Treefrog's ***call is a soothing, melodic trill*** as opposed to the harsher, faster, metallic trill of the Cope's. Other differences, though less reliable, include the Gray Treefrog having ***skin that's more warty and black edging around blotches***. The tadpoles of both species have high tail fins tinged with red or orange, and the newly metamorphosed treefrogs are bright green.

Voice: Soft, melodic trills.

A recent metamorph that shows the tadpole tail being absorbed.

Gray Treefrog male singing its melodic trill. Note yellow inner legs.

Gray Treefrogs may be gray or green. Treefrog color is mostly a response to its environment, for camouflage, but humidity and temperature are also considered factors. The black edges of blotches remain, whether the frog is gray or green. Juveniles are mostly all green.

Courtship, Mating & Eggs: Gray Treefrogs breed in the middle of May, their soft melodic trills coming from ponds and marshes. The males call from the water or from perches on vegetation protruding from the water or hanging over it. Males hold territories about 30 inches from another male until they can attract a female and breed. Then a satellite male will likely take over the perch. The female lays 1000 to 2000 eggs as the male fertilizes them. The eggs are attached in small clusters to vegetation near the water's surface. Tadpoles hatch in three to seven days. The little bright green frogs leave the pond in six to eight weeks.

Gray Treefrogs will hunt on the sides of buildings and on windows, clinging with suction pads on the toes.

Range Notes: Manitoba and Minnesota south into Texas and northeast into Maine. Broad areas of overlap with Cope's; both species sometimes calling from the same ponds.

Spring Peeper

Pseudacris crucifer

frogs

Bogs, small ponds and other wetlands in deciduous and coniferous forests.

Nature Notes:

Spring Peepers overwinter hidden under leaf litter, logs, or other material. They freeze, but glycerol in their body turns to glucose and acts as antifreeze. Its distribution protects organs and muscles, while allowing ice between them. With the spring thaw, the peepers also thaw for an early migration to the breeding pond.

Mating pair in amplexus.

This tiny frog with loud whistling peeps is one of the earliest to call.

Description: Length is ¾ to 1 ¼ inch.

The Spring Peeper is a treefrog, with small toe pads. Its color can be tan, brown, or gray, and ***usually has a characteristic dark X on its back***, although the X is often crooked or with extra lines.

Voice: A whistling peep that slurs slightly up at the end.

Courtship, Mating & Eggs: Spring Peepers are among the earliest breeding frogs, calling from temporary ponds, bogs, and other shallow wetlands in late March or early April. The advertisement call is a whistling peep that slurs slightly up at the end. A male peeps about once per second. From a distance, a chorus of peepers sounds similar to sleigh bells. Spring

Spring Peeper atop a frozen pond. They freeze solid in winter but thaw in spring.

Eggs are attached to submerged vegetation singly or in small groups.

Perched on a Pitcher Plant cup, a recent metamorph shows the tadpole tail being absorbed.

The "X" on the back doesn't always look like an X.

Peeper males call from perches on vegetation sticking out of the water or along the shore and maintain territories. The Spring Peeper warning call is a stuttering *"creeek"* that sounds very similar to a chorus frog. Satellite males keep a low profile, looking to intercept a female or replace a successful resident male. The female lays 750 to 1300 eggs which are stuck to submerged vegetation singly or in small groups. The eggs hatch in a week or two and the tadpoles transform in 45 to 90 days.

Range Notes: North American range from central Manitoba to Quebec and south to east Texas and northern Florida.

frogs

Boreal Chorus Frog

Pseudacris maculata

Marshy or boggy areas of forests and open woodlands.

Nature Notes:

The Boreal Chorus Frog seems poorly equipped for survival. Its short tibia provides less leverage, reducing jumping ability to mere hops. The toe pads of chorus frogs are small, limiting climbing ability.

The Boreal Chorus Frog can be greenish, reddish gray or brown.

A tiny early calling frog that sounds like a fingernail running across the teeth of a comb

Description: Length is ¾ to 1½ inches.

The Boreal Chorus Frog, until recently, was considered a subspecies of the Western Chorus Frog. The Boreal Chorus Frog is often reddish brown, gray or greenish, and ***on its back are three green stripes that break into spots. Sometimes it looks more spotted than striped.*** The tibia (leg between thigh and foot) is shorter. But these two frogs are so similar they were long confused.

The best way to differentiate is to note where it was found, as DNA studies have shown that Boreal Chorus Frogs are the chorus frogs found in Minnesota, Wisconsin, and Isle Royale.

Voice: Very similar to the clicking "*crrreeek*" call of the Western Chorus Frog.

The Boreal Chorus Frog often looks more spotted than striped (see Western Chorus Frog on page 78).

There is much variation in pattern and color.

Singing begins very early in the spring.

Courtship, Mating & Eggs: Boreal Chorus Frogs hibernate close to the pond to get the early jump on breeding. The call is very similar to the clicking *"crrreeek"* call of the Western Chorus Frog. Eggs are laid in several loose clusters, hatch in a few days, and the tadpoles may take two or three months to metamorphose.

Range Notes: In Minnesota and Wisconsin, the Boreal Chorus Frog is now found throughout both states following mitochondrial DNA studies. In Michigan, the Boreal Chorus Frog is found only on Isle Royale. The Western Chorus Frog's range has been retracted to just lower Michigan.

Found from northeast Alberta to Hudson Bay and south through Illinois and west with extensions into Idaho, Utah and Arizona. A separate population is in southwest Quebec.

Western Chorus Frog

Pseudacris triseriata

frogs

Marshes, swamps, or damp woods.

Nature Notes:

Chorus frogs are hard to find because of their small size and cryptic coloration. They often call while hidden in grassy hummocks or other vegetation and usually go silent when approached. After the breeding season, they are dispersed and rarely encountered.

A tiny frog that hibernates next to a pond, so it can be one of the first to call.

Description: Length is ¾ to 1½ inches.

The Western Chorus Frog is tiny. The slightly bumpy skin is brown, tan, gray, or olive, with *three dark stripes down the back*. Occasionally the stripes break up posteriorly or are absent. A dark stripe runs through the eye, and a white stripe runs along the upper lip. The toe pads are small. Western Chorus Frogs are so similar to Boreal Chorus Frogs, the key is where it was found. *If the chorus frog was found in Lower Michigan, then it's a Western Chorus Frog.*

Voice: A rising *"crrreeek"* sound, similar to running a thumbnail along the teeth of a comb.

Courtship, Mating & Eggs: Chorus frogs are the first frogs to call in spring, usually beat-

The inflated vocal pouch acts as a resonator, amplifying the call. In general, the smaller the frog, the higher the pitch of the call.

ing spring peepers by a day or two. They cheat. They hibernate under logs or thick grasses close to the pond. The Western Chorus Frog call sounds like the call of the Boreal Chorus Frog. Males call for about a week before the females arrive. During amplexus the female lays 500 to 1500 eggs, attaching elongated loose clusters to submerged plants. They hatch in 3-14 days and metamorphose 40 to 90 days after that.

Range Notes: Mitochondrial DNA studies of chorus frogs have found that the Western Chorus Frog's range is greatly reduced and it is now only found in Lower Michigan in our area. Minnesota, Wisconsin and Isle Royale in Lake Superior have the Boreal Chorus Frog. The issue is not settled yet.

North American range from lower Michigan south into north central Tennessee and northeast around Lake Erie to western New York.

American Bullfrog

Lithobates catesbeiana

Permanent water with lots of vegetation, such as ponds, lakes, river sloughs or backwaters.

Nature Notes:

American Bullfrogs are voracious eaters that don't limit themselves to invertebrates. They have been known to eat fish, other frogs including bullfrogs, small turtles, birds, and mammals.

Often a bullfrog gives a warning call as it avoids danger by leaping into the water. A captured bullfrog may scream like a cat. The sudden shriek might startle a predator into dropping it. It might also attract a bigger predator to distract the first one.

Our biggest frog with a voracious appetite, devouring most any critter that will fit in its mouth.

Description: Length is 3½ to 6 inches.

The American Bullfrog has ***a ridge, the tympanic fold, which curves over and behind the tympanum. It does not have dorsolateral ridges.*** The webbing of its foot does not reach the end of its longest toe. The male's tympanum is much larger than its eye, while the female's is about equal in size. The male's throat is yellow.

Voice: The male's call is a bass toned, vibrating "*jug-o'-rum*" or "*rrrruuuummm*" that's loudest in the middle. The vocalization is produced as the internal vocal pouch swells the throat.

Courtship, Mating & Eggs: In our area American Bullfrogs breed during June and July. The male is territorial, sometimes having to win a shoving match to keep other males 6

The female's tympanum is the same size as her eye (left). Males have a tympanum larger than the eye (right).

A bullfrog lurks beside a wetland shore.

Bullfrog eggs float black side up to absorb sunlight and warmth.

Bullfrog habitat along backwaters of the Mississippi River

to 18 feet away. The female chooses the male and lays 5,000 to 20,000 black and white eggs during amplexus. The eggs float, black side up, in a thin sheet for 3 to 6 days until the tadpoles hatch. The tadpoles will overwinter once or twice, by hibernating in the bottom mud, before they metamorphose the following summer.

Range Notes: Southern Quebec to central Florida and west through Iowa and Nebraska to Texas and northern Mexico. Then New Mexico, southern Arizona, California and north into British Columbia.

Green Frog

Lithobates clamitans

frogs

Shallow water and along the shores of most lakes and other permanent slow-moving or still bodies of water.

Nature Notes:

The Green Frog is the most commonly seen frog in most of our area. It often sits along the shore, waiting for an insect snack.

Sensing danger, the Green Frog gives a squeaky warning cry while leaping into the water.

This frog's call is the explosive *"gunk!"* like a loose banjo string emanating from a wetland.

Description: Length is 2¼ to 3½ inches.

The Green Frog can be various shades of swamp green or brown with ***dark crossbands on the legs***. A dorsolateral ridge extends at least half way down the body. The ***webbing of the hind foot is incomplete, with most of the longest toe free***. The mature male has a yellow throat, and its tympanum is larger than the eye.

Voice: The male's advertisement call is an explosive *"gunk!"* like a loose banjo string. Often the call is instantly repeated with decreasing volume, resembling an echo. A warning call, like a louder single note advertisement call, is used to hold territory, but a physical battle is sometimes necessary.

A male Green Frog hunts prey along the shore.

Mature male revealing its bright yellow throat.

Even if the entire frog is mostly brown, the upper lip will still be green.

Females lay 3,000 to 4,000 eggs.

Dark bands cross the Green Frogs legs, while the Mink Frog has bands that run lengthwise on the leg.

Courtship, Mating & Eggs: Green Frogs breed from late May into August. The female Green Frog chooses a male (or a good territory) and lays 3,000 to 4,000 floating eggs during amplexus. They hatch in 3 to 5 days. The tadpoles may overwinter before metamorphosing.

Range Notes: Where their ranges overlap, Green Frogs and Mink Frogs avoid competition with Green Frogs close to the shore and Mink Frogs usually out in deeper water areas.

The Green Frog is found in southern Quebec west into southern Ontario and south through eastern Minnesota to east Texas and east to the Atlantic Ocean and northern Florida.

Pickerel Frog

Lithobates palustris

Grassy areas adjacent to clean, clear, cool water, such as trout streams, spring-fed bogs, fens, and other wetlands.

Nature Notes:

Pickerel Frogs have toxic skin. Many predators that eat frogs, such as gartersnakes (*Thamnophis*) or watersnakes (*Nerodia*) usually won't touch the Pickerel Frog. Herons apparently recognize Pickerel Frogs and won't eat them. Pickerel Frogs are devoured by Mink, *Mustela vison*, and by Bullfrogs and Green Frogs. A Pickerel Frog's skin secretions may kill another frog, including another Pickerel Frog, if placed in the same collecting bag.

The frog with squarish brown spots, yellow inside the thighs and groin area and toxic skin

Description: Length is 1 ¾ to 3 inches.

The Pickerel Frog is tan or brown with ***two rows of roughly square darker brown spots down the back***, flanked by prominent dorsolateral ridges. Spots elsewhere on the body and legs also appear roughly square or rectangular. When the hind legs are folded into their ready-to-jump position, the black bands on them often line up.

The groin and concealed surfaces of the hind legs are washed with bright yellow.

Voice: The male's call is a soft one to two second snore, done while floating in the water or completely submerged.

Courtship, Mating & Eggs: Pickerel Frogs breed in April in the clear, cool, unpolluted

Note the roughly square brown spots.

The spots aren't very square but the yellow in the groin identifies this as a Pickerel Frog.

A large clump of the Pickerel Frog's eggs attached to a submerged stick.

water they inhabit. The 800 to 3,000 eggs, brown above and yellow below, are attached in a loose clump to submerged vegetation. Tadpoles hatch in 10 to 21 days and metamorphose in two to three months.

Range Notes: From our area they range northeast past Ontario and Maine to Nova Scotia and south to Missouri and east Texas.

Northern Leopard Frog

Lithobates pipiens

Grassy areas near water. Lakes are needed for hibernation. Shallow ponds and backwaters are used for breeding.

Nature Notes:

The Northern Leopard Frog's preference of grassy areas for foraging often leads them to the large lawns of businesses. During dry weather, the frogs take advantage of automatic sprinklers for moisture, as they snap up the insects attracted by security lights.

Rarely you may find abnormally pigmented frogs such as the albino (bottom photo).

The frog named for its distinct round black spots.

Description: Length is 2 to 3½ inches.

The Northern Leopard Frog is green or brown with ***black "leopard" spots. Dorsolateral ridges are prominent and extend to the groin.***

Voice: The advertisement call is a long snore, which lasts several seconds and may be followed by some short clucks. The release call, given by males and unreceptive females, is a chuckling sound.

Courtship, Mating & Eggs: The Northern Leopard Frog breeds in early spring, although not as early as the Wood Frog, peepers, or chorus frogs. Before the ice is gone from their lake or other large body of water, the Northern Leopard Frogs head out and migrate to marshes, ponds, or backwaters to breed. These shallow

Northern Leopard Frogs have twin vocal pouches located above the front legs. Their call is a long snore.

Mass of hundreds of eggs. Mating takes place in April and May.

Leopard Frog coloration is variable; they can be either green or brown. But they always show their "leopard spots."

waters are warmer. The male Northern Leopard Frog vocalizes while floating in the water or sometimes while submerged. During amplexus, the female lays 300 to 6500 eggs that are black on top, white below, and attached as an oval mass to submerged plants. Sometimes egg masses from many females are found in the same part of the pond.

Range: Quebec and New England west to Saskatchewan and south to Arizona, Iowa and Pennsylvania.

Pesticides in agricultural runoff are suspected in the deformities of frogs found is such areas. This frog has grown two extra legs.

Mink Frog

Lithobates septentrionalis

Lakes that have lots of floating or emergent vegetation. Sphagnum bogs bordering lakes.

Nature Notes:

As could be guessed by their fully webbed feet, Mink Frogs are more aquatic than other frogs. They habitually hang out in deeper water than other frogs and their diet includes a bigger percentage of aquatic invertebrates.

When they do venture out onto a sphagnum mat, they're usually within jumping distance of water.

Mink Frogs often sit on floating lily pads, sometimes distant from shore.

Tadpoles may take two years to become adults.

If you scratch its skin, this frog smells like a mink.

Description: Length is 1¾ to 2¾ inches.

The Mink Frog, usually looks similar to the Green Frog, but there are several differences. The Mink Frog has ***fully webbed feet***, with webbing out to the last joint of the long toe and to the end of the outside toe. The Mink Frog is usually ***heavily mottled with dark spots or blotches***, but more importantly, it has ***large dark spots, or sometimes lengthwise bars or stripes, on its hind legs***. Green Frogs have some dark crossbands on the hind legs. Dorsolateral ridges vary from absent to prominent.

If the Mink Frog's skin is scratched, it gives off an odor like a Mink or like rotten onions. The male has a tympanum larger than the eye, and often a yellow throat.

Voice: The male's call, made while floating in

The Mink Frog's large dark hindleg blotches is a good field mark to look for.

Mink Frogs prefer northern lakes with floating vegetation.

Note their fully webbed feet.

the water or perched on a lily pad, is a rapid knocking *"cut cut cut cut."* It sounds like distant hammering. A chorus sounds like galloping horses on a cobblestone street.

Courtship, Mating & Eggs: Mink Frogs usually breed in June and July. The female attaches 2,000 to 4,000 eggs to the stem of a plant, often deeper than three feet below the surface of the water. The tadpoles, because of the cold northern lakes and short growing season, may take one or two years to metamorphose.

Range Notes: Where their ranges coincide, Mink Frogs and Green Frogs avoid competition with the Mink Frogs usually out in deeper water areas and the Green Frogs hanging out near the shore.

Ranges from northeast Minnesota and Manitoba northeast and east to Quebec and Labrador and into northern New York.

Wood Frog

Lithobates sylvatica

frogs

Woods; deciduous, coniferous or mixed. A closed canopy forest retains enough moisture for them. Only enters water for breeding.

Nature Notes:

The Wood Frog hibernates in the woods, under logs or leaf litter or loose soil. Here it is subjected to freezing temperatures, but survives by producing large amounts of glucose, which serves as an antifreeze. The organs and muscles are protected while allowing ice to form in spaces between them.

Wood Frogs in the northern part of our states often have a white stripe down the back (Portage County, Wisconsin).

This frog can be various shades of brown with a black mask across its eyes.

Description: Length is 1 ½ to 2 ¾ inches.

Recognized by its ***dark mask***, the Wood Frog can be tan, or various shades of brown tinted with gray, red, pink, or bronze. The dark brown mask starts at the snout, runs across most of the eye, covers the tympanum, and ends above the front leg. It effectively hides the eye, which is very dark to match the stripe. A white stripe runs along the upper lip. The dorsolateral ridges may be accentuated by small black spots. Wood Frogs from northern Minnesota and Upper Michigan may have a white middorsal stripe (see sidebar photo).

Voice: A rapid-fired *"rack, rack, rack, rack"* is similar to a quacking duck.

Courtship, Mating & Eggs: Wood Frogs are one of the earliest frogs to breed in spring,

This frog almost escaped notice by hiding in fallen leaves.

Calling male's vocal pouches are located above the front legs.

Wood Frog eggs compared to Leopard Frog eggs.

Female Wood Frogs often lay their eggs in a sunny corner of a pond allowing sunlight to warm the eggs and speed egg development.

usually in a temporary woodland pond. The male floats on the surface while vocalizing His thin vocal pouches, located over the front legs, expand with each *"rack."* The female lays from 500 to over 3,000 eggs which are attached just below the surface to a submerged plant. The Wood Frogs disperse back into the woods after 6 to 14 days. The tadpoles hatch in four days to four weeks, depending on the highly variable temperatures of early spring.

Range Notes: The Wood Frog's range extends farther north than any other North American amphibian and is the only amphibian found north of the Arctic Circle.

Ranges from Alaska across Canada into Quebec and south to scattered populations in Georgia and Arkansas.

Order Testudines: Turtles

The turtle is characterized by its protective shell. The bony structure of the shell is the rib cage, in which greatly expanded ribs and the vertebrae are fused together. Overlying the bone are horny shields called scutes or plates, although a few turtles just have a thick, leathery covering. The shoulders and hips have evolved to positions inside the rib cage. The top part of the shell is called the carapace, the bottom is the plastron, and they are connected on each side by the bridges. The plastron of some males, depending on species, is noticeably concave to fit over the female's carapace during copulation. Most turtles can withdraw the head and legs into the shell for protection. Those species with less armor due to a reduced plastron compensate with aggressive biting or speed. Really!

Another trait of turtles is having no teeth. They do have a sharp, horny beak covering the jaws. Turtles can't chew, but they bite into food and tear it to pieces with their clawed feet. Most turtles need to be submerged to swallow.

Turtles have sharp vision and most are wary.

Male turtles have a longer, thicker tail with the vent located farther out. The males of some species have long front claws.

All turtles are oviparous, laying shelled eggs, usually in a nest dug in the ground by the female. She covers the eggs and abandons them. Predators find most nests and eat the eggs. Those eggs that escape detection hatch in roughly 7 to 13 weeks, depending on species and incubation temperature.

That isn't all temperature affects. Some turtles have temperature-dependent sex determination. It seems strange, but there is a critical time during the middle third of incubation when the egg temperature determines the sex of the turtle. At higher temperatures, only females are produced. At lower temperatures, only males are produced (Painted Turtle, the three map turtles, Blanding's Turtle, Spotted Turtle and Eastern Box Turtle). Two more species (Snapping Turtle and Eastern Musk Turtle) fit this scenario, except that at still cooler temperatures they produce females again. Genetics seems to determine the sex ratios of Spiny Softshells and Wood Turtles, even though they don't have sex chromosomes like birds and mammals.

Cooler temperatures and shorter summers place limits on the reproduction of our northern states turtles. Farther south many turtles nest twice a summer. Some of our Painted Turtles nest twice, but most, along with other turtle species, only nest once a season. It also takes more years for some species to reach maturity in our area.

There are about 263 species of turtles worldwide, divided into about 13 families. The fourteen species of turtles in our states represent four families.

Family Chelydridae – Snapping Turtles

Snapping turtles are brown or very dark, with a large head and long tail. The plastron is small, giving the legs mobility but not protection. Big sharp jaws and a nasty disposition compensate for the reduced plastron. One species, the Snapping Turtle, is found in our area.

Family Kinosternidae – Musk & Mud Turtles

Musk and mud turtles are small brown or black turtles with large heads and short tails. The elongated carapace is high and the plastron is small. A musky odor is emitted from glands on their sides where the skin meets the bridge. Musk and mud turtles are poor swimmers, preferring to walk on the bottom. Except for nesting, they don't often leave the water. When they are on land, they are aggressive biters. One species, the Eastern Musk Turtle, inhabits waters in our area.

Family Emydidae – Pond, Marsh & Box Turtles

This large family includes the typical aquatic turtles, and also the terrestrial box turtles. The former are often seen basking on logs, ready to drop into the water if approached.

Ten species in seven different genera are found in our northern states: Painted, Spotted, Blanding's, Wood, Northern Map, Ouachita Map, False Map, Eastern Box and Ornate Box Turtles plus the Pond Slider.

Family Trionychidae – Softshells

Softshell turtles are highly aquatic and built for a submerged lifestyle, which explains their unique appearance. The flattened shell has a tough, leathery skin instead of horny scutes. The plastron is much reduced posteriorly, allowing free movement of the rear legs for swimming. The feet are large and fully webbed and not encumbered with scales. The neck is long, and the nose is like a snorkel. Two species, the Smooth Softshell and the Spiny Softshell, swim in our rivers and lakes.

Turtle Hatchlings

Northern Map Turtle (pg 110)

Ouachita Map Turtle (pg 112)

Northern False Map Turtle (pg 114)

Woodland Box Turtle (pg 116)

Plains Box Turtle (pg 118)

Red-eared Slider (pg 120)

Midland Smooth Softshell (pg 122)

Eastern Spiny Softshell (pg 124)

Snapping Turtle

Chelydra serpentina

Most permanent bodies of water, such as lakes, rivers, swamps, or ponds. Mud bottoms and aquatic vegetation are preferred.

Nature Notes:

A large Snapping Turtle on land is a formidable sight. The reduced plastron allows it to walk with the body well off the ground. It drops down and turns with gaping jaws to face any threat. The snapper sometimes raises its posterior and lunges forward as it strikes out with its long neck.

In recent years snapping turtle numbers have declined, and the sizes of snappers harvested for food have also declined. Some regulations have been tightened.

Rare two-headed hatchling.

This is the biggest and certainly the heaviest turtle in our area, and it will bite.

Description: Carapace length is 8 to 14 inches. Record is 19 3/8 inches.

Mature Snapping Turtle is ***huge with a large head with big jaws, a long neck, and a long, spiked tail.*** The carapace scutes are raised and pointed at the rear, forming three keels. But the keels smooth out as the turtle ages. The color above is brown to very dark brown.The plastron is small and cross-shaped, leaving the legs exposed from below.

Hatchlings have a very rough carapace and a tail equal to or longer than the shell.

Diet: A wide variety of invertebrates, small vertebrates and even plants. Studies, even one done in a waterfowl sanctuary (Lagler, 1943), have shown that Snapping Turtles are NOT a serious threat to waterfowl. In fact, the few

Over 100 eggs may be laid in a nest excavated in soft soil (normal is 25 to 50). Nests are vulnerable and often raided by raccoons, fox, skunk and mink.

The tiny plastron is little protection. Fortunately, Snappers have few enemies and other formidable defenses. Note the long spiked tail.

ducklings lost to snappers are insignificant compared to the nest raids of mammalian predators.

Courtship, Mating & Eggs: Sometimes two males will battle over the right to mate with a female. Courtship involves swimming face to face with necks extended and swinging back and forth.

A female snapper might trek the better part of a mile to find a suitable nesting site. Over 100 spherical white eggs might be laid, but usually 25 to 50. Incubation time ranges from 55 to 125 days, but most hatch in two months. Some stay in the nest over winter and emerge the following spring. Hatchlings are always at great risk. Some have a long journey to water and may not get that far before dehydration kills them. Others will be eaten, either before or after reaching the water. Once the hatchlings enter the water, some may drown, because juvenile snappers are poor swimmers.

Females might take 17 to 20 years to reach maturity

Range Notes: Found across the U.S. east of the Rockies.

Eastern Musk Turtle (Stinkpot)

Sternotherus odoratus

Shallow, clear water in lakes or slow-moving streams or backwaters.

Nature Notes:

Eastern Musk Turtles are lousy swimmers and prefer to walk on the bottom in shallow water. When still, they look more like rocks than turtles. Sometimes they bask out of water and have a reputation for climbing out on overhanging trees and dropping into a passing boat.

Apparently, they are preyed upon by the Red-shouldered Hawk. I saw a photo of one flying with an Eastern Musk Turtle in its talons.

Although the plastron has a hinge, the plastron is too small for the hinge to be effective.

A small, frequently overlooked turtle, because it's on the bottom of the stream.

Description: Carapace length is 2 to 4½ inches. Record is 5 7/8 inches.

The Eastern Musk Turtle was named because of its musky odor, and for a while, it was called "Stinkpot." It has a ***domed, smooth elongated carapace*** but the plastron is small, offering little protection. The seam between the second and third sets of plastral scutes is hinged. Except for ***two yellow stripes on the side of its pointed head, one above and one below the eye***, its skin color is drab brown to black. It has barbels on its chin and neck. The male's tail is short with a hardened nail at the tip.

Hatchlings are black with tiny white spots around the edge of the carapace and have the

The Eastern Musk Turtle's yellow head stripes are key to identification.

Their plastron is small offering little protection.

Hatchlings are nearly black with small light spots around edge of carapace.

yellow head stripes. The carapace has at least a mid-dorsal keel; sometimes two more keels.

Diet: Aquatic insects, snails, amphibian larvae, earthworms, crayfish, fish and aquatic vegetation.

Courtship, Mating & Eggs: The Eastern Musk Turtle is the only turtle in our three states that might not bother digging that flask-shaped nest for the eggs. Sometimes the female lays the eggs, usually 3 to 5, in a shallow excavation or under the edge of a log or bush. They might be half covered or not covered. The eggshells are thick and brittle which helps to keep them from drying out. They hatch in 60 to 90 days.

Range Notes: Along the Mississippi in southwestern Wisconsin and up the Wisconsin River. Southern Lower Michigan.

Ranges northeast into Canada over Lake Erie and Lake Ontario, then south through Ohio to central Texas and east to all of Florida and then northeast to Maine.

Western Painted Turtle (subspecies of Painted Turtle) *Chrysemys picta bellii*

Slow-moving streams, ponds, lakes, and swamps with aquatic vegetation, muddy bottoms and basking sites.

Nature Notes:

Western Painted Turtles are the most common and most obvious chelonian in Minnesota and Wisconsin. They're especially noticeable in the early spring when the sun is warm, but the water is cold. In ponds where logs are few, turtles can be lined up or stacked on every available surface.

These eggs will hatch in 50 to 80 days...if not dug up and eaten by raccoon, fox, skunk or other predators.

The most common turtle in Minnesota and Wisconsin; most noticeable when basking on a sunny April day.

Description: Carapace length is 4 to 7 inches.

The Western Painted Turtle is one of two subspecies found in our three states. It has a low, smooth, dark olive-green carapace with extensive red coloring on the plastron and ventral surface of the marginals. The upper jaw has a pair of sharp cusps flanking a notch. It has yellow stripes but *lacks the red stripes on the legs and neck*. The plastron has a larger dark pattern that extends out on some of the seams.

Some Painted Turtles have a dark reddish-brown coating on the shell, most visible on the plastron. This is caused by absorbing chemicals in the water. When the outer layer of the scutes is shed, so is the color. Males have long front claws and are smaller than females.

Over 35 painted turtles vie for space on a prime basking log.

The hatchling's plastron is very colorful.

This turtle just came out of hibernation, because the dried mud is still on his shell.

The brightly colored hatchlings have a more circular carapace with a slight vertebral keel.

Diet: Omnivorous, feeding on aquatic plants, insects, worms, minnows, tadpoles and more.

Courtship, Mating & Eggs: The male swims in front of the female, extends his front feet and flutters his long claws on the sides of her face and neck. She will usually reach out and stroke his front legs. She then sinks to the bottom and he follows and mounts her from behind.

Females lay 3 to 20, more likely 6 to 9, elliptical white eggs. The eggs hatch in 50 to 80 days and the young often overwinter in the nest. Females take 6 to 10 years to reach maturity while males can breed in 3 to 7 years.

Range Notes: Ranges north into Canada, west through the Dakotas and south into southern Missouri and Kansas. In eastern Wisconsin and in the middle of the Upper Peninsula, it intergrades with the Midland subspecies.

Midland Painted Turtle (subspecies of Painted Turtle) *Chrysemys picta marginata*

Lakes, ponds, slow moving rivers, river backwaters, marshes and other permanent waters. Need mud bottom and basking sites.

Nature Notes:

Although Painted Turtles need to be submerged to eat, I've seen them crawl out of the water to snatch prey off the bank and return to the water to swallow it.

Because Painted Turtles are wide ranging and abundant in many areas, they often serve as a child's introduction to reptiles.

I remember catching my first painted turtle out of a small stream when I was in fourth grade.

The most common turtle in Michigan, and especially noticeable when basking on a sunny April day.

Description: Carapace length is 4 to 7 inches.

The Midland Painted Turtle is one of two subspecies found in our area. Its carapace is smooth, low and has no keel. ***Yellow and red stripes are on the forelegs and neck.*** The plastron is dull yellow, sometimes tinged with red or orange and mostly unmarked except for a large dark blotch in the middle.

Females are larger than males, and males have long front claws.

Hatchlings have a round carapace, and they do have a keel. Their plastron color is more vivid than adults.

Diet: Midland Painted Turtles are omnivorous, feeding on both animals and plants. Worms, tadpoles, aquatic insects and fish are

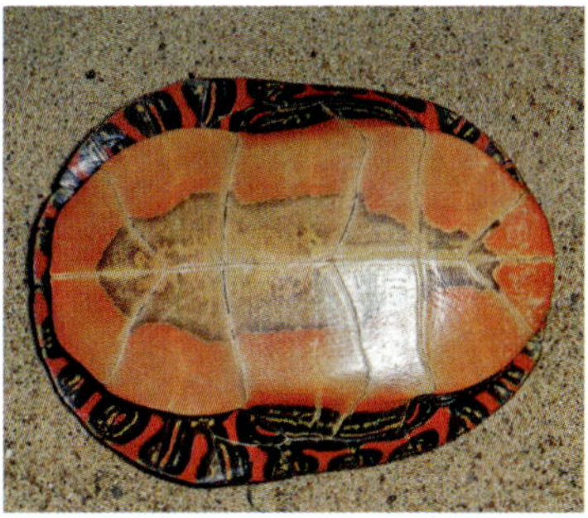

Compare plastron pattern to Western Painted Turtle. Note how the dark area of the plastron is concentrated in the middle.

Red-striped legs and red lower neck striping help separate this subspecies from the Western Painted Turtle.

Juveniles are nearly miniature versions of the adults, but note the hatchling's keeled shell and their more vivid plastron coloration.

on the menu as well as duckweed and other aquatic plants.

Courtship, Mating & Eggs: The male swims in front of the female and vibrates his long front claws against her head and neck. She sometimes responds by reaching out and stroking his front legs with her own claws. When she sinks to the bottom, he follows and grips her shell from behind and they mate. June is peak nesting time. The female digs a flask-shaped hole with her hind feet, sometimes moistening the soil for easier digging, and deposits 3 to 20 eggs. She covers and abandons her nest. Predators like skunks and raccoons will search. Eggs that escape detection will hatch in 50 to 80 days, but sometimes the young don't emerge until spring.

Range Notes: Found in the eastern Michigan U.P. and Ontario into Quebec and western Vermont and southwest into western Kentucky and southern Illinois. Another subspecies, the Eastern Painted Turtle, is found from Alabama to North Carolina and up the coast to Labrador.

Spotted Turtle

Clemmys guttata

Small ponds, shallow marshes, bogs, and swamps. They need a mud bottom and vegetation.

Nature Notes:

In early spring, Spotted Turtles, especially males, move through deciduous woods and enter other ponds, including small vernal pools. On sunny days they will bask on logs, muskrat houses or just grass hummocks. If disturbed while basking, they drop into the water and hide at the bottom under leaves or other debris.

During summer, Spotted Turtles are more secretive and may spend a lot of time just burrowed into the mud.

A Spotted Turtle survived at least 26 years in the wild and a captive lived 42 years.

A small gem of a turtle, it's attractively decorated with bright yellow spots.

Description: Carapace length is 3½ to 4½ inches.

The Spotted Turtle has a ***smooth black carapace sprinkled with yellow spots.*** The ***head is black with yellow or orange spots.*** The male has dark jaws and brown eyes, while the female has light jaws, a yellow or orange chin and orange eyes.

The carapace of a hatchling Spotted Turtle usually has a maximum of only a single spot in every large scute, but sometimes none.

Diet: Insects, worms, snails, crayfish, tadpoles, and aquatic vegetation.

Courtship, Mating & Eggs: In early spring, male Spotted Turtles will battle each other prior to courtship and mating with a female. They bite and try to overturn each other until

Captive turtle eating a minnow.

Note the bright yellow spots on the head.

Plastron is distinctively marked.

The light colored upper jaw show that this is a female.

one is driven away. The winning male will bite at the female's back legs and tail and the edges of her carapace. Then he mounts her from behind and bites at her head and neck while hanging on to her carapace with his claws. He slips backwards or to one side to get his tail under hers and mate with her.

This hatchling has only one spot per scute.

In June, the female digs a nest with her hind feet and lays 2 to 7 eggs. If undetected by raccoons, skunks and a host of other predators, they will hatch in 45 to 83 days.

Spotted Turtles take 7 to 14 years to reach breeding age.

Range Notes: From southern Michigan the Spotted Turtle ranges around the eastern Great Lakes and scattered populations just south. Separated range covers states along the east coast.

Blanding's Turtle

Emydoidea blandingii

Shallow marshes, bogs, swamps, and river backwaters. Aquatic vegetation and a muddy bottom are important.

Nature Notes:

The crayfish is the preferred food item and is stalked underwater. The Blanding's Turtle strikes with a sudden thrust of its head and long neck. Its jaws open and the neck expands to suck in both water and prey. The crayfish is seized in the sharp jaws, torn apart and devoured.

Plastron has black patches on outer edges.

A big turtle with a bright yellow chin and throat and a long neck.

Description: Carapace length is 6 to 9 inches. Record is 11 1/8 inches.

The Blanding's Turtle has a ***long neck with a bright yellow chin and throat that contrast sharply with the black or brown color of its shell, head, and legs***. The jaw is notched in front, giving the appearance of a smiling turtle. The ***smooth, domed carapace, which is shaped much like a helmet, is sprinkled with yellow or tan***. The plastron is yellowish with black patches at the outer rear of each scute. The plastron is hinged, allowing the two halves to be pulled against the carapace to protect the head and legs.

Hatchlings have a flatter carapace and the plastron is yellow with one central dark blotch. The plastral hinge is not yet developed. The tail is about the same length as the hatchling's shell.

Easily identified by its long neck and bright yellow chin and throat.

Plastron detail showing growth annuli.

Blanding's have deeply notched upper jaw.

Diet: Crayfish and other invertebrates, tadpoles, frogs, fish and plants.

Courtship, Mating & Eggs: The male Blanding's Turtle swims after a female, approaches her from behind and moves over the top of her shell. He bites at her head and neck. With his long neck stretched out, he swings his head from side to side in front of her until she allows him to mate with her. The female usually digs her nest in an open site with sandy soil and lays 6 to 21 elliptical eggs. They hatch in 50 to 75 days.

Blanding's Turtles take 14 to 20 years to reach maturity.

Studies of some populations have found that each year less than half of the adult females lay eggs. But as females age, they nest more often and have larger clutches of eggs. They seem not to decline physically and those over fifty years old are more productive than ever.

Range Notes: From our region into central and southern Iowa and east into Ontario and western Quebec. Isolated populations scattered including in Nebraska, southeast New Hampshire and eastern Massachusetts.

Wood Turtle

Glyptemys insculpta

At home in woods and water. Sandy-bottomed streams and rivers that flow through deciduous woodlands.

turtles

Nature Notes:

Wood Turtles have a unique method of finding worms. The turtle alternates stomping one front foot on the ground and then the other. Or the turtle lifts and drops the front of the plastron onto the ground. Possibly the sound or vibrations resemble raindrops, and the earthworm emerges to avoid drowning only to be gobbled up by the hungry Wood Turtle. Wood Turtles don't wander more than a few hundred yards from their streams, so home ranges of up to a mile necessarily follows streams and rivers.

Sandy habitats preferred.

This aptly named turtle has a shell that appears to be carved from wood...and it is found in wooded habitats.

Description: Carapace length is 5 ½ to 8 inches. Record is 9 7/8 inches.

The Wood Turtle's ***carapace has the appearance of sculptured wood.*** The scutes are brown, sometimes with radiating light or dark streaks, and each has ***concentric growth rings rising to form low pyramids. From these high points, tiny ridges radiate outward.*** The head is black, while the skin from the neck onto the legs is yellow or orange. The front legs have sturdy scales for better protection.

Hatchling Wood Turtles are plain brown or gray with a weakly keeled carapace and long tail.

Diet: Various plants like dandelions, raspberries and blackberries plus invertebrates like

Ideal habitat for a Wood Turtle: A sand-bottomed river with heavily wooded shores.

The carapace appears sculpted. This feature helps differentiate from Blanding's Turtle.

Dark blotches are on rear outer edges of the plastron.

Two hatchling Wood Turtles emerge from their nest along a wooded river.

earthworms, slugs and insects. They can eat while on land and often forage there.

Note the Wood Turtle's yellow throat and upper legs.

Courtship, Mating & Eggs: Both male and female Wood Turtle extend their heads out towards each other, lower their heads and move them from side to side. Mating is usually in shallow water but can be on land as well. In June, the female will find a nesting site on a high sand bar, sandy riverbank or terrace. The female digs a flask-shaped hole and deposits 3 to 18 elliptical eggs. She uses her back feet to fill the nest and often uses her plastron to smooth it over. If the nest is not sniffed out by a raccoon or other predator, the eggs will hatch 58 to 71 days later.

Range Notes: Our region to Ontario and the Northeast including Quebec, Maine and Labrador down through Pennsylvania to northern Virginia.

Northern Map Turtle

Graptemys geographica

Rivers. They may also be found in backwaters, side channels, reservoirs, and lakes. Fallen trees are favorite basking sites.

Nature Notes:

The males and females of this river turtle have different sized heads which gives them different diets. Given the different feeding habits of the males and females, they are not competing for the same food. That allows for a higher population of turtles.

Like other map turtles, the Northern Map Turtle spends much time basking in the sun on logs or muskrat houses. It is wary and quick to drop into the water if someone approaches.

A turtle mainly found in rivers.

Description: Carapace length is 3 ½ to 6 ¼ inches (male) and 7 to 10 ¾ inches (female).

The Northern Map Turtle's identifying head marking, to distinguish it from the other two species of map turtles in our area, is ***a single, isolated yellow spot behind the eye. There are no other noticeable spots or blotches.*** The carapace has a brown or greenish gray often with dark blotches. There is a vertebral keel, and the rear of the carapace is serrated.

The plastron is straw-colored or cream-colored. Besides their size difference, females have a lower keel and a large broad head. The male has a smaller, narrower head, long front claws and a thick tail.

Young individuals have light map-like lines on the carapace, hence the name. Hatchlings have a gray to brown keeled carapace and dark

The Northern Map Turtle's upper jaw is curved. Also note the fine yellow and green striping on the neck.

This is a female as indicated by her large broad head.

The plastron is creamy yellow in color. Carapace is serrated at the rear.

Females take 10 to 14 years to reach egg-laying maturity.

markings along the plastron seams.

Diet: Northern Map Turtles are unusual in that males and females have significantly different diets. The female's large jaws have broad crushing surfaces adapted for crunching through the shells of clams, snails, and the exoskeleton of crayfish. The males feed on insects and smaller snails and crustaceans. Dead fish are scavenged by both sexes. All feeding is done while submerged.

Courtship, Mating & Eggs: The female lays 6 to 20 elliptical eggs during June or early July. They hatch in 50 to 70 days. Males are ready to breed in three to five years, but the larger females need 10 to 14 years to reach maturity.

Range Notes: Distribution follows rivers in Minnesota and Wisconsin: Mississippi, Minnesota, St. Croix, Wisconsin and Wolf rivers. Widespread in lower Missouri and lower Arkansas to northern Mississippi. Found in Ontario and into Quebec. Follows rivers in New York and Pennsylvania.

Ouachita Map Turtle

Graptemys ouachitensis

Rivers. Lakes and large ponds when they are close to a river. Basking sites, such as large fallen trees are important.

Nature Notes:

Like other map turtles, Ouachita Map Turtles spend hours basking in the sun, perched on logs or muskrat houses. They are alert and wary and quick to drop into the water if disturbed.

They seek shelter from the river current behind submerged logs or wing dams for sleeping or hibernating along with other map turtles.

For a few years, this turtle was called a Southern Map Turtle. Now it's considered a full species and is again called Ouachita Map Turtle.

Common name is pronounced "*Wash' i taw*" or alternatively "*Wah cheeta*."

A locally common turtle of the Mississippi, Minnesota and Wisconsin Rivers.

Description: Carapace length is 3½ to 6½ inches (male) and 4¾ to 10¼ inches (female).

Of the Ouachita Map Turtle's ***identifying bold yellow head spots, the largest is behind the eye, with another below the eye and a third spot at the rear of the lower jaw.*** An unpaired spot is at the front of the lower jaw. The carapace is grayish brown or olive with dark blotches (about one per each large scute). It has a prominent vertebral keel. Females do not have an enlarged head. Males have long front claws and a thick tail.

Hatchlings are round with more noticeable head markings.

Diet: Insects, snails and aquatic plants.

Courtship, Mating & Eggs: The male swims in front of the female and vibrates his

Note the adult's diagnostic head spots.

Notice the shedding outer layer of scutes to make room for further shell growth.

Crawling out of its leathery eggshell.

Note the hatchling's extensive face spots.

long front claws on the side of the female's head. He gets behind her and grips her carapace from behind and they mate. As soon as late May or early June, the female will find a nesting site on a sandbar, beach, or sandy bank and dig a cavity with her hind feet. She lays 8 to 17 leathery eggs, covers them and abandons them. She may lay a second clutch of eggs in late June or early July.

In September, hatchlings emerge from the early clutch nests. Hatchlings from late clutch nests won't emerge until spring.

Range Notes: In our area along the Mississippi, Black and Wisconsin rivers. It is not found in Michigan.

Follows rivers to the Gulf, including in Kansas and Oklahoma. Wide ranging in northern Arkansas and Louisiana. Follows rivers into Tennessee and Indiana with scattered populations in Kentucky and Indiana.

Northern False Map Turtle (subspecies of False Map Turtle) *Graptemys pseudogeographica pseudogeographica*

Large rivers. Lakes and large ponds if they are close to a large river.

turtles

Nature Notes:

False Map Turtles are often seen sunning on fallen trees or logs in rivers. They are wary and quick to drop off into the water if approached.

Fallen trees and logs, as well as wing dams, offer shelter from the river current while they sleep or become inactive during winter.

In Kansas, Missouri and central Illinois the Northern False Map Turtles are intergrades with the Mississippi Map Turtle, the other subspecies, which is found all the way to the Gulf of Mexico.

An uncommon turtle of the Mississippi and Wisconsin rivers.

Description: Length is 3 ½ to 5 7/8 inches (male) and 4 ¾ to 10 7/8 (female).

The Northern False Map Turtle found in our area is a subspecies of the False Map Turtle. It is best identified by head markings to distinguish it from the other two species of map turtles in our three states. ***It has a small bar, sometimes a narrow crescent shape, behind the eye. There are no other large blotches or spots on the head. All lines and stripes are narrow and at least three stripes from the neck go all the way to the eye.*** The carapace is brown or olive colored and has a vertebral keel. Often each large carapace scute has a dark blotch. Females do not have an enlarged head. Males have long front claws and a thick tail.

Hatchlings are round and have a prominent vertebral keel.

Hatching out of the leathery egg shell.

A recent hatchling in shallow water.

Adult basking in the sun.

A male turtle with his long front claws and thick tail.

Diet: Eat while submerged, feeding on various aquatic invertebrates: insects, insect larvae, snails, clams, crayfish and vegetation.

Courtship, Mating & Eggs: The male uses his long front claws to vibrate against the female's head before mating with her. In late May or June, she digs a cavity with her hind feet in a sand bar or sandy bank near the water and lays 12 to 22 eggs. She covers them and crawls away. Some females come ashore in July and lay a second clutch of eggs. If not devoured by predators, the eggs hatch in 60 to 75 days. If the young haven't emerged by the middle of September, they can wait until spring.

The hatchlings are round with a strong vertebral keel. The tiny white egg tooth on the turtle's snout, which cut through the eggshell, will fall off in a few days.

Range Notes: In our region follows big rivers: Mississippi, Minnesota and Wisconsin.

North Dakota in the Missouri River and down the river to Missouri.

Woodland Box Turtle (subspecies of Eastern Box Turtle) *Terrapene carolina carolina*

Deciduous or mixed forests, forest edges and semi-open areas near shallow water, such as bogs, marsh edges or small streams.

turtles

Nature Notes:

Woodland Box Turtles are terrestrial. They forage on land for worms, slugs, snails, berries, and a variety of other food items.

Their behavior is greatly influenced by weather. They often slog around in boggy areas during hot days. During extended heat and drought, they dig burrows and aestivate (remain dormant) until conditions improve.

They are most active during and after rain.

Box turtles retreat into burrows for hibernation, and they have the option of digging farther as the weather gets colder.

A terrestrial turtle, it needs the extra protection of a high shell that closes tightly.

Description: Carapace length is 4 ½ to 6 inches. Record is 9 ¼ inches.

The Woodland Box Turtle is the subspecies found in our area. It has a ***highly domed carapace***, and the plastron is hinged, allowing it to be pulled snug and tight against the carapace when more protection is needed. The shell color varies. It can be brown or very dark with yellow or orange streaks, but the yellow or orange can also be the dominant color. Usually the male has red eyes and the female has brown eyes.

Hatchlings have a flatter shell but a distinct sometimes yellow keel. The hinge is not yet functional.

Diet: Earthworms, slugs, snails, insects and berries.

Shell color and markings vary from dominated by yellow (left) to mostly brown with some yellow spots (right).

A hinged plastron allows for full closure and safety from predators.

Juvenile showing flatter shell with vertebral ridge (above) and more colorful plastron.

Courtship, Mating & Eggs: Mating seems an impressive accomplishment for a male box turtle. When mounting the female from behind, he has to overcome the female's high domed carapace, and the fact that both male and female have short tails. But the posterior third of the male's plastron is concave, and he hangs onto the female's carapace with the specially curved claws on his rear legs. The male's plastron is vertical while mating. Afterwards, the male often tips over backwards, risking death if he cannot right himself. The female selects a nesting site, usually an area open to the sun, and deposits 3 to 11 eggs. The eggs hatch in 50 to 90 days, and the young may overwinter in the nest. Box turtles need about ten years to reach maturity.

Range Notes: From southern Michigan it's found southwest through southern Illinois and southeast through Georgia and northeast into the southern tip of Maine. The other subspecies, the Three-toed Box Turtle, is found in southern Missouri to eastern Texas and east into Alabama.

Plains Box Turtle (subspecies of Ornate Box Turtle)

Terrapene ornata ornata

Dry, sandy prairies and oak savannas.

Nature Notes:

Plains Box Turtles are especially out and about after a heavy summer rain.

Hibernation is by burrowing into the sandy soil, usually on a southwest facing slope. As it gets colder, they can burrow deeper, although some have been known to endure freezing temperatures for several weeks.

Juveniles do not yet have the high domed carapace.

A turtle of the prairie, it needs the extra protection of a high shell that it can tightly close.

Description: Carapace length is 4 to 5 inches. Record is 6 1/8 inches.

The Plains Box Turtle is the subspecies of the Ornate Box Turtle found in our area. The ***shell is high domed but flat on the top***. The carapace is dark but streaked with yellow. The plastron is brown and also streaked with yellow. A hinge runs across the plastron between the front third and the rear two thirds, allowing the plastron to be closed tight against the carapace and giving the terrestrial little turtle the ultimate protection against predators.

Males have yellow, orange or red spots on their head and legs. They usually have a red iris, and the plastron is concave. Females are spotted with yellow and have a brown iris. Their plastron is flat or convex.

Shell is domed but flat on top (compare to Woodland Box Turtle). Patterns vary in individuals.

Plastron is hinged to shut tight. Male's plastron is concave and female's is convex.

Males have red eyes and often show red spots on the legs.

Hatchlings are flatter and have a yellow vertebral stripe. The hinge is not yet functional.

Diet: Spring food includes earthworms, beetles, caterpillars and prickly pear cactus. Summer adds mulberries, strawberries, raspberries and blackberries.

Courtship, Mating & Eggs: Most breeding activity is in the spring but can be in any month the turtles are active. The male nips at the female's carapace and then mounts her from behind, helped by his concave plastron. He has specially curved claws on the first toes of his back feet to grip her shell. While mating, the male often slips backwards but his rear feet hang on and they also prevent the female from closing her shell. In late May or June, the female will dig a hole with her back feet, lay 2 to 8 eggs and cover them with the sandy soil. They will hatch 59 to 70 days.

Range Notes: Found along the lower Wisconsin River and a few other isolated spots. Also Iowa, southwestern Illinois and then widens its range from South Dakota to southern Texas and extreme southwest Wyoming, eastern Colorado.

Red-eared Slider (subspecies of Pond Slider)

Trachemys scripta elegans

Ponds, lakes, river backwaters and marshes with basking sites, aquatic vegetation and ittle or no current.

Nature Notes:

This is the turtle that was sold in dime stores decades ago. This practice was finally ended to prevent children from getting Salmonella from handling the turtle.

Red-eared Sliders frequently bask in the sun, but they are alert and wary. The common name "slider" comes from disturbed turtles quickly sliding off their logs and dropping into the water.

Hatchling Red-eared Slider.

A turtle of the Southeast; the scattered populations in our area were likely all introduced or escaped pets.

Description: Carapace length is 5 to 8 inches. Record is 11 7/8 inches.

The Red-eared Slider is the subspecies of Pond Slider found in our area. Well named for the ***distinct red bar that extends from behind the eye to over its ear.*** The weakly keeled carapace is olive brown with yellow and black vertical streaks. The head and limbs have thin yellow stripes.

Males have long front claws. Old males often turn dark and the pigment obscures colors including the red bar on the head.

The plastron is yellow with black spots or black circles, usually in the middle of every large scute.

Juveniles have a stronger keel and green carapace.

The red "ear" field mark can be seen from quite a distance. This mark is often darker and can be obscured in old males.

Round dark blotches usually mark each yellow scute on the plastron.

The bold red bar behind the eye is the "ear" in Red-eared Slider.

Diet: Aquatic insects, snails, crayfish, tadpoles, frogs and fish are on the menu with aquatic vegetation.

Courtship, Mating & Eggs: The male swims in front of the female, extends his front legs and vibrates his front claws against the side of her head. She might swim away, but the male is persistent and keeps getting in front of her and vibrating his front claws. He finally mounts her from behind and mating is accomplished. In late May or June, she digs a flask-shaped cavity in the soil and deposits 2 to 30 eggs. She covers the eggs and crawls away. The eggs are on their own and most are found and devoured by predators, such as raccoons, skunks, mink, foxes, and more. Eggs in undetected nests incubate for 65 to 80 days before they hatch. Sometimes the hatchling turtles remain in the nest until spring.

Range Notes: This is a southern turtle. It inhabits all three of our three states in small, isolated populations, most or all of which, were introduced.

Midland Smooth Softshell (subspecies of Smooth Softshell) *Apalone mutica mutica*

Rivers and large streams with sandy bottoms. Mud bottoms are also used.

turtles

Nature Notes:

The Midland Smooth Softshell is highly adapted for life in the water. Its streamlined shape, unhindered rear legs and large fully webbed feet allow for swift movement through the water. It frequently buries itself in sand so that its long neck and snorkel-like nose can reach the surface. The softshell can also absorb oxygen through the pharyngeal linings of its throat, its cloacal lining and its skin, all of which helps the softshell remain underwater for long periods of time.

When basking on a riverbank, they face the water for a quick escape.

A turtle of sandy bottomed streams where it burrows to hide or ambush a meal.

Description: Carapace length is 4½ to 10½ inches (male) and 6½ to 14 inches (female).

The Midland Smooth Softshell is the subspecies of Smooth Softshell found in our area. Its pancake-like ***shell is smooth, flat, flexible and lacks any horny plates***. It has a long tubular nose with ***no ridges inside the nostrils***. The neck is long, and the feet are large and strongly webbed. Two lines, peach or cream colored and edged by black, are on the side of the head. One starts under the nostrils and the other behind the eye. The plastron is white or light gray and is short at the rear, so it doesn't interfere with movement of the rear legs.

Males have a light brown carapace with small, scattered dark brown or black markings. The tail is thick. Females have dark brown blotches on their carapace, and they have a smaller tail.

Softshells often hide under the sand with only the head sticking out.

Compare Spiny (left) to Smooth (right).

Smooth Softshell just hatching.

A basking Smooth Softshell displays its long neck.

Diet: Insects, mollusks, crustaceans and fish.

Courtship, Mating & Eggs: The male searches for a receptive female and then swims over the top of her while they mate. In June or early July, she finds a nesting site near the water and digs a cavity with her hind feet. Four to 33 spherical, hard-shelled eggs are deposited and covered. Eggs hatch in 56 to 84 days. The hatchling's egg tooth is not used, but instead its front claws rip open the eggshell.

Range Notes: Found in the Mississippi, Minnesota and Wisconsin rivers.

Found in major rivers from extreme southern North Dakota to Kansas and through Missouri, Iowa and Illinois and a few isolated places like Ohio and western Pennsylvania.

Nature Notes:

Softshell turtles are actually fast and on land they scamper like rabbits.

If grabbed, a Midland Smooth Softshell usually will not bite but just retreat into its shell.

Eastern Spiny Softshell (subspecies of Spiny Softshell) *Apalone spinifera spinifera*

Rivers with sandy bottom and sandy banks. Will also inhabit lakes and waters with muddy bottoms.

Nature Notes:

Softshells are highly adapted to life in water. Their streamlined shape, unhindered rear legs, and large fully webbed feet allow swift movement through the water. A softshell often buries itself in sand, so its long neck and snorkel-like nose can reach the surface. The softshell also can absorb oxygen through the pharyngeal linings of its throat, the cloacal lining, and its skin, all of which help the turtle remain submerged for long periods of time.

When sunning on a riverbank, the turtle faces the water for a quick escape.

This softshell is actually speedy on land.

When grabbed, Spiny Softshells bite aggressively.

Built for speed, this turtle is occasionally spotted by canoeists as it flies by like a submerged frisbee.

Description: Carapace length is 5 to 12 3/16 inches (male) and 7 to 21¼ inches (female).

The Eastern Spiny Softshell is the subspecies of Spiny Softshell found in our area. It bears a definite resemblance to a large pancake. Its shell is round, flattened and lacks horny scutes. The edges are flexible. The plastron is reduced posteriorly and doesn't hinder the hind legs. The legs and feet are not scaled, and the feet are large and strongly webbed. The softshell's neck is long and its nose is long and tubular There are two black-bordered yellow stripes, one behind the eye and one behind the mouth.

Adult ***males have small spots and circles called ocelli on their shell***, and the ***carapace feels gritty like sandpaper***. Adult females are larger with a blotchy camouflage pattern. ***At the front***

Note the C-shaped nostrils that identify this species. The snorkel-like nose helps reach the surface while the turtle remains hidden. The namesake spines are at the front edge of female's carapace (top right).

Hatchling emerging from its egg.

Softshells will often bury themselves in sandy bottoms.

edge of the female's carapace are sharp, spiny projections. Each *nostril is C-shaped* because of a horizontal extension inside (top left photo).

Diet: Prefers crayfish and other aquatic invertebrates but would not turn up its long nose at a tadpole or fish.

Courtship, Mating & Eggs: Softshell claws are sharp and the shell is vulnerable, so the male hovers over the female during mating. Softshells will nest as close to the water as possible. She selects a site on a raised sandbar or on the slope of a riverbank to lay her eggs. She quickly digs a hole and deposits 9 to 38 spherical white eggs. She covers them and returns to the safety of the water. The eggs hatch in 55 to 85 days. Males reach maturity in 4 to 5 years, while the larger female requires 8 to 10 years.

Range Notes: East into Ontario and northeastern New York, south into northern Alabama and northwest Mississippi, and west to Oklahoma, Kansas and Nebraska. Found farther in isolated spots and rivers from eastern Massachusetts to central Montana.

Order Squamata: Lizards & Snakes

Lizards and snakes are covered with scaly skin that protects the body and retains moisture. Some lizards and snakes have smooth scales and others have keeled scales. A keeled scale has a central ridge running its length, and the overall effect is that the reptile has a rough appearance. Smooth scales make an animal look shiny. The type of scale a lizard or snake has is important in identification. The outer layer of skin is periodically replaced as the animal grows.

Keeled scales

Smooth scales

The tongue is used for sensory purposes as it flicks out and picks up minute particles from the air. Some lizards have a long forked tongue like a snake, while most have a shorter, blunt-tipped tongue. The tongue tips contact Jacobson's organ, located on the roof of the mouth. The sense, similar to the sense of smell, is called chemoreception, and is useful in detection of food, enemies, or the opposite sex.

Male lizards and snakes have a double penis called hemipenes. Only one is used during mating.

Suborder Sauria – Lizards

Even though a few lizards are legless, they differ from snakes in several ways. Most lizards have eyelids while snakes never do. Lizards have ear openings and snakes don't. The lower jaw bones of lizards are fused, so they can't stretch apart to accommodate large prey. Lizards do not have the broad ventral (belly) scales that snakes can use for crawling. Some lizards can break off the tail to escape a predator and grow a new one. About 4,450 species of lizards in 22 families are known worldwide. Four species in three different families are found in our northern states.

Family Anguidae – Glass Lizards

The glass lizards are legless lizards that differ from snakes. They also have a tiny osteoderm or bony plate inside every scale. This gives them a stiff, hardened body that will not coil up like a snake. They get a little flexibility from the groove that runs low on both sides and will extend a little to accommodate carrying developing eggs or a large meal. Although they look smooth and shiny like glass, their name comes from sometimes breaking their long tail into multiple pieces if grabbed by an enemy. The pieces twist and turn and distract the potential predator, giving the lizard a chance to escape and grow a new tail.

One species, the Slender Glass Lizard, inhabits our area.

Family Scincidae – Skinks

Most skinks are designed for escape. Often a fast moving skink doesn't use its small legs, instead folding them back and using a rapid snake-like swimming motion. If the tail is grabbed, it promptly breaks off, leaving the predator with a wiggling distraction. The skink's tail will regenerate, though it will be dull colored with no pattern. If the skink is caught, its streamlined shape and smooth, shiny scales, make the squirming lizard slick and difficult to hold.

The Five-lined Skink and Prairie Skink are found in our area.

Family Teiidae – Racerunners

A racerunner has different kinds of scales with tiny, granular scales on its back and sides, larger rectangular scales on its belly and large scales on top of its head. Keeled scales are on the tail. It has strong rear legs for running. Racerunners are easily distinguished from the smooth, shiny, streamlined looking skinks.

Granular scales

In our area there is one species: the Six-lined Racerunner.

Suborder Serpentes – Snakes

Snakes have no legs, no eyelids, and no ear openings. The eye is protected by a clear scale called the brille. The lower jaw bones are not fused together but are connected by stretchable ligaments, allowing the jaws to engulf large prey. The jaws alternate as each side reaches forward and pulls the prey a little farther into the mouth. Tiny backward-curved teeth help maintain a grip on the animal.

Snakes move by lateral undulation, which is the side to side motion used for swimming, or crawling fast through bushes or on the ground. The snake pushes off from S-shaped loops of the body. Snakes can slowly crawl straight ahead by rectilinear locomotion, using the broad belly scales or ventral scutes, as each scale edge pushes against the ground, shifts forward and pushes again. Some scales are doing the gripping and pushing while others are sliding ahead. This technique works well enough that some snakes can use it to crawl up a tree trunk

Prior to shedding, lymphatic fluid moves between the new skin and the outer layer. The eyes turn blue but then clear just before shedding. The snake rubs its nose against objects to start pulling the skin off. The snake crawls out of the skin as it peels back and is left inside out.

To eat, some snakes simply seize the prey in their jaws and overpower it. Tiny, sharp teeth curve backwards, helping to secure the animal. Some snakes grasp the prey, wrap coils around it and constrict tight enough to stop the animal's heart or breathing. A rattlesnake injects venom as it strikes the prey and then avoids further risk by letting the animal wander off to die. The venom starts the digestive process. The rattlesnake locates the animal by following its scent and swallows it head first.

Family Colubridae – Typical Snakes

The family Colubridae is a huge conglomeration of harmless snakes, containing about 1,560 species throughout the world. In North America, any snake that isn't a blind snake, boa, sea snake, coral snake, or pit viper was dumped into this family. Colubrids are a diverse group of snakes. Some have rear fangs with mildly toxic saliva, some constrict their prey, and some lay eggs while others bear live young. Many of these harmless snakes imitate rattlesnakes by vibrating their tails, which often makes a buzzing sound against dry leaves or brush. Colubridae is already divided into subfamilies. In the future one or more of these could be elevated to the family level.

Our region is home to 23 species of these so-called typical snakes.

Family Crotalidae – Pit Vipers

The Copperhead, Cottonmouth, and rattlesnakes are all pit vipers. Their eyes have an elliptical pupil, which appears as a vertical slit in daylight but expands wide in the dark of night. But the real adaptation to night is the pair of facial pits, for which they were named. Each pit is located between the nostril and the eye. They detect heat. Specifically, they detect infrared radiation which allows the snake to accurately strike a mouse in complete darkness. Pit vipers are venomous snakes with a pair of large hollow fangs that fold back against the roof of the mouth when not in use. The fangs are connected by ducts to the venom glands, which are modified saliva glands at the rear of the head. The size of the venom glands gives the pit viper its characteristic triangular-shaped head. When the snake strikes, the mouth opens wide and the fangs rotate out, so they point forward. As the fangs stab into the target, muscles contract to force venom out through the fangs. The snake controls the amount of venom injected. It may not inject any venom if the strike is defensive.

Two pit vipers, the Timber Rattlesnake and the Massasauga, which is also a rattlesnake, inhabit parts of our three states.

Timber Rattlesnakes are found in the river bluff country of southeast Minnesota and southwest Wisconsin.

Upper Great Lakes Snake Habitats

This old field has sandy soil, making it ideal for Eastern Hog-nosed Snakes, Smooth Greensnakes and Eastern Massasaugas. Look on overcast days or early morning sunlight.

Bluff prairies in SW WI often host Prairie Ring-necked Snakes, Blue Racers and Timber Rattlesnakes. Overcast or early morning is best. The ring-necks will be under flat rocks.

Baxter's Hollow in Sauk County, WI, is a large deciduous woods with a creek and is a good place to look for Eastern Hog-nosed Snakes, greensnakes, milksnakes or foxsnakes.

An area of sand dunes in Wabasha County, MN has both species of hog-nosed snakes as well as the Bullsnake. When it gets too hot they all go underground so look early.

Oak savanna in Monroe County, WI has some excellent oak savanna on sandy soil. Eastern Hog-nosed Snakes and Bullsnakes are found here. You might hear one of these snakes hissing before you see it.

An old railroad bed running through the Tiffany Wildlife Area in Buffalo County, WI, provides a place for foxsnakes and massasaugas to sun themselves. Best time is after a rain when the sun comes out.

The Chippewa River bottoms flooding in Buffalo County, WI, is a good place to find foxsnakes and massasaugas, especially when the floodwaters are still rising. Best is to walk around the edge which might have lots of displaced snakes.

A Mississippi River bluff in Crawford County, WI, has exposed rocks and a bluff prairie. Look for Prairie Ring-necked Snakes, Eastern Milksnakes, Blue Racers and Timber Rattlesnakes. The ring-necks and milksnakes will be under flat rocks.

Western Slender Glass Lizard (subspecies of Slender Glass Lizard) *Ophisaurus attenuatus attenuatus*

Oak savannahs and sand prairies.

Nature Notes:

About two thirds of the glass lizard's length is tail. In the tail are multiple fracture zones or places where the tail can break. This lizard is fast, but if a predator grabs the tail, it will break off easily and can break into multiple pieces. The wiggling tail distracts the predator and the lizard will likely escape. The lizard will grow a new tail, although it will never be as long as the original and will be a darker color with no pattern (photo below).

Regenerating tails often lack the stripes of original tail.

lizards

This is a strange beast that looks like a snake, but in reality it's a legless lizard.

Description: Length is 22 to 42 inches. Maximum snout-vent length is 11 3/8 inches.

The subspecies found in our area is the Western Slender Glass Lizard. It is ***a legless lizard and snake-like in appearance***. It differs in having eyelids and ear openings. Also, low along the side of the body is a lateral groove which offers some flexibility for the body to expand to accommodate a large meal or developing eggs. Less obvious is the fact that the lower jaw bones are fused and won't stretch to allow the lizard to swallow bigger prey. It lacks the broad belly scales of a snake which are used to crawl straight ahead. Instead the lizard whips back and forth, pushing off on loops of its body. The body is stiff and the lizard doesn't have the flexibility to coil up like a snake.

Though it has no legs, the Glass Lizard is not a snake.

Hatching Glass Lizard. Juveniles are identical to adults in color and markings.

Sand prairies and oak savannahs are the preferred habitat.

Ear holes and eyelids help identify this reptile as a lizard, and not a snake.

Stripes extend down the entire body and tail, with a narrow brown stripe down the middle of the back and a wider black stripe along each side. Hatchling glass lizards are miniature replicas of the adults.

Diet: Mostly invertebrates like crickets, grasshoppers, beetles and spiders. Also, eggs of ground-nesting birds and small vertebrates.

Courtship, Mating & Eggs: The male grips the female's neck in his jaws and they mate. She hollows out a space under a rock or log and lays 5 to 17 eggs in June or July. She stays with her eggs until they hatch in about 50 to 60 days.

Range Notes: Only found in south central Wisconsin in our area. Also northwest Indiana and southern Iowa. Missouri and Kansas to the Gulf.

Ideal Glass Lizard habitat in southcentral Wisconsin—sandy oak savannah.

lizards

Common Five-lined Skink

Plestiodon fasciatus

Damp sandy soil with logs or rocks for basking. Forest edges, oak or pine barrens, rock outcrops, and beach areas with driftwood.

Nature Notes:

This diurnal skink becomes active in late April or early May. They warm themselves by basking in the sun, usually on a log or rock.

When threatened, the skink may run up a tree trunk, but more likely it will tuck its small legs along its sides and rapidly wiggle snake-like into cover. It may also dash into its burrow, which it previously dug in the sandy soil.

If grabbed by the tail, the skink can break it off, leaving the predator with a wiggling distraction. The skink grows a new tail, which will be a dull gray and never as long (below).

lizards

The female of this little lizard species will tend her eggs until they hatch.

Description: Length is 5 to 8 ½ inches long. Its maximum snout-vent length (since the tail may be lost or shorter due to regeneration) is 3 3/8 inches long.

The Common Five-lined Skink is a ***smooth shiny lizard with small legs***. Adult males are ***brown with faint stripes*** and a gray tail. Males have orange jaws which become reddish orange during the breeding season. A ***young adult female has five thin white or yellow stripes against a black or brown background, with the dorsal stripe splitting into a "Y" on the head.*** The tail is bright metallic blue. Older females look faded with tan stripes on a brown background and a blue-gray tail.

Juveniles are colored like young females and have more intense colors.

Female skinks guard their eggs until they hatch. When the tiny blue-tailed young emerge they quickly disperse.

Recently emerged juveniles.

Female guarding her offspring.

Diet: Mostly invertebrates. Occasionally, a large skink will eat a baby mouse or tiny frog.

Courtship, Mating & Eggs: The male defends territory, allowing the presence of blue-tailed juveniles and females but chasing away orange-faced males. About a month after breeding, the female lays 5 to 18 eggs in a moist, hidden place. She tends the eggs and guards them for 30 to 60 days until they hatch, and the young disperse. Skinks mature in the second or third year.

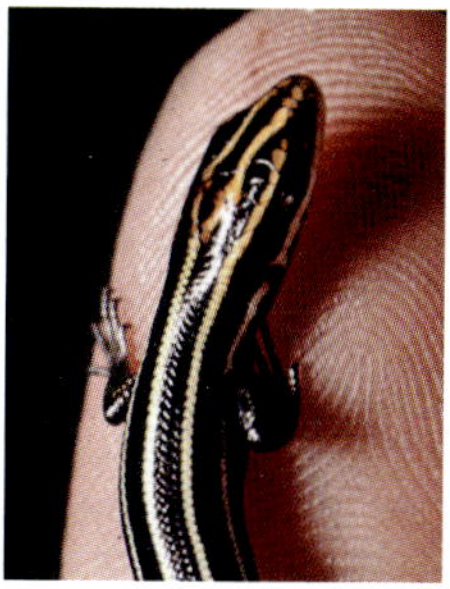

Note Y-shaped top stripe.

Range Notes: Our region to Ontario and southwest New York, south to north Florida and west to east Texas.

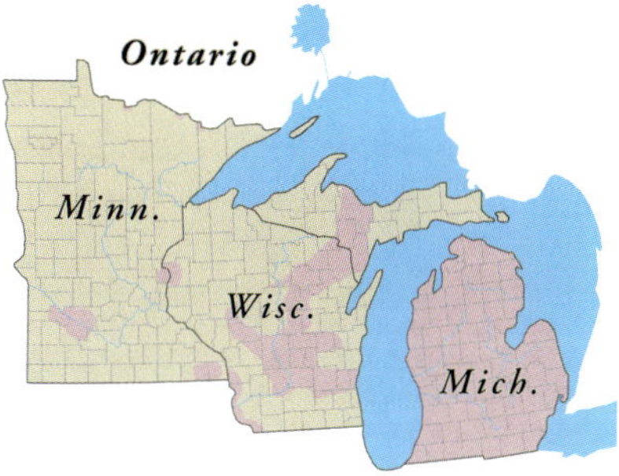

Northern Prairie Skink (subspecies of Prairie Skink)

Plestiodon septentrionalis septentrionalis

Sandy soil habitats; pine barrens, steam banks, road cuts, or grassy areas. Also, rock outcrops and gravelly glacial outwashes.

Nature Notes:

If grabbed, the tail breaks in one of the special fracture zones and the skink's blood vessels contract to minimize blood loss. The tail will regenerate, but it will be shorter and plain looking.

Prairie Skinks dig burrows in the sandy soil where they spend the night or where they can retreat during the hottest part of the day. When they do come out, they bask in the sun and hunt invertebrates.

This female is moving her eggs down the burrow since the rock over them was removed.

Smooth, shiny and streamlined, this little lizard is fast but can drop its tail if a predator grabs it.

Description: Length is 5 ¼ to 8 ¾ inches including tail. Maximum snout-vent length (since the tail may be lost or shorter due to regeneration) is 3 ½ inches long.

The Northern Prairie Skink is the subspecies of Prairie Skink found in our area. ***Smooth and shiny with small legs***, it's gray to light brown with stripes. The ***most prominent stripe is a thin white or light dorsolateral stripe bordered by wider black or very dark brown stripes.*** On its back are two dark stripes which can be either distinct or faded looking. Adult males develop orange faces during the breeding season.

Juvenile Prairie Skinks have seven light stripes and metallic blue tail.

lizards

This Prairie Skink's orange face means that this is a male (inset). During the breeding season the orange becomes even brighter.

Habitat for both Prairie Skinks & Five-lined Skinks in the Minnesota River valley

Diet: Invertebrates such as crickets, grasshoppers, beetles, caterpillars and spiders.

Juveniles have metallic blue tails.

Courtship, Mating & Eggs: In the spring, male skinks develop orange coloration around their mouth and throat. They hold a female with their jaws gripping her neck and mate with her. In late June or early July, she hollows out a nesting site and lays 5 to 13 eggs. She tends the eggs and guards them until they hatch in 40 to 52 days. She will stay and guard the young for two or three days until they disperse. The young are ready to breed during their second spring.

Range Notes: From our area west and south including the eastern edge of the Dakotas, half of Iowa, parts of Nebraska and Kansas.

Prairie Racerunner (subspecies of Six-lined Racerunner)

Aspidoscelis sexlineata viridis

Bluff prairies, sandstone outcrops, dry prairies, sand outwashes in river floodplains and Mississippi River dikes.

lizards

Nature Notes:

Well named. On a warm day they can run up to 18 mph.

Racerunners are fast getting out of the egg too. After photographing hatching turtles, skinks and snakes, I was fascinated by how fast these little guys emerged.

Racerunners dig their own burrows in the sandy soil. There they hibernate, estivate, escape a predator, spend the night or spend any day not warm enough for their satisfaction.

Male's throat becomes blue during the breeding season.

A fast little lizard, these guys are tough to catch once they get warmed up.

Description: Length is 6 to 10½ inches including the tail. Maximum snout-vent length is 3 1/8 inches.

The subspecies of Six-lined Racerunner found in our area is the Prairie Racerunner. Similar in size to the skinks, ***racerunners do not have a smooth, shiny, streamlined body shape***. They have a conglomeration of different sized scales with large scales on the top of the head, tiny granular scales on the legs and body, large rectangular scales on the belly and large keeled scales on the tail. ***None of them are shiny.***

The head has a pointed snout, and the rear legs are bigger than the front legs.

A brown dorsal stripe is flanked by three yellow or greenish stripes on each side. Males are light blue on the sides of their faces and bright

Racerunners are ready to run as soon as they leave their eggshell (left). Juveniles are similarly colored to non-breeding adults but have bluish-green tails (right).

Sandy, rocky dikes along the Mississippi River can be excellent habitat for Prairie Racerunners.

yellowish-green on the anterior of their bodies during the breeding season. Juveniles are similar in color to non-breeding adults but have light blue tails.

Diet: Crickets, grasshoppers, katydids, beetles, moths, spiders.

Courtship, Mating & Eggs: The male displays his bright colors but ends up chasing the female anyway. He grabs her by the neck, and they mate. The female digs a cavity in the sand or sandy soil and lays 1 to 8 eggs. Unlike the skinks, she abandons her eggs. I could find no reference to how many days it takes for the eggs to hatch, so I checked my notebook. I had a Prairie Racerunner from Houston County, Minnesota that laid five eggs on June 24, 2006. They all hatched on August 16; 53 days later.

Range Notes: On or near river bluffs in our region. Widespread from southwest Missouri to the Texas Gulf and then northwest to Nebraska and Colorado.

Western Wormsnake

Carphophis vermis

Bluff prairies and semi-wooded bluffs. Needs rocks to hide under.

Nature Notes:

Western Wormsnakes are very secretive, rarely appearing above ground. They spend most of their life under rocks or in burrows, most likely to emerge on a rainy summer night.

The Western Wormsnake has enlarged rear teeth, but it is not known if the saliva is toxic which could help subdue their prey. It is no threat to people, as it hardly ever bites and its head and mouth are too small to be effective against a person.

The species name "*vermis*" means worm and supposedly describes the snake's body.

snakes

This little snake is at the northern tip of its range, and due to its secretive nature, is hardly ever seen.

Description: Length is 7½ to 11 inches. Record is 15 7/8 inches.

The Western Worm snake is a unique looking two-toned snake of contrasting halves. The *top half of the snake is dark brownish black or grayish black, sometimes with a purple tint. The bottom half of the snake is a bright pinkish or salmon color.* The scales are smooth. The pointed *head is small and so are the eyes.* Its *tail ends in a sharp, hard point or spine.*

A hatchling has the same colors and pattern as the adult.

Diet: Earthworms. Rarely soft insect larvae, and one ate a baby ring-necked snake.

Courtship, Mating & Eggs: Western Worm snakes are known to mate in the fall and in

A very small and secretive snake that is rarely seen in the wild.

Hatchlings emerge about 50 days after the eggs are laid. They are miniature versions of the adults.

the spring. At that time, they are on southwest facing bluffs where their hibernacula are located, and considering their small size, that makes it easier for them to find each other. The male and female do find each other and mate. Later, the eggs are reported to sometimes be visible through her translucent belly skin. She lays 1 to 12 eggs underneath rocks or inside rotting logs or in burrows, and they hatch in about 50 days.

Range Notes: In our three states there is only one isolated population: in Grant County, Wisconsin, about 120 miles north of the next closest population. Scattered in southern Iowa. Missouri and northwest half of Arkansas west into southeast Nebraska, Kansas and Oklahoma.

Kirtland's Snake

Clonophis kirtlandii

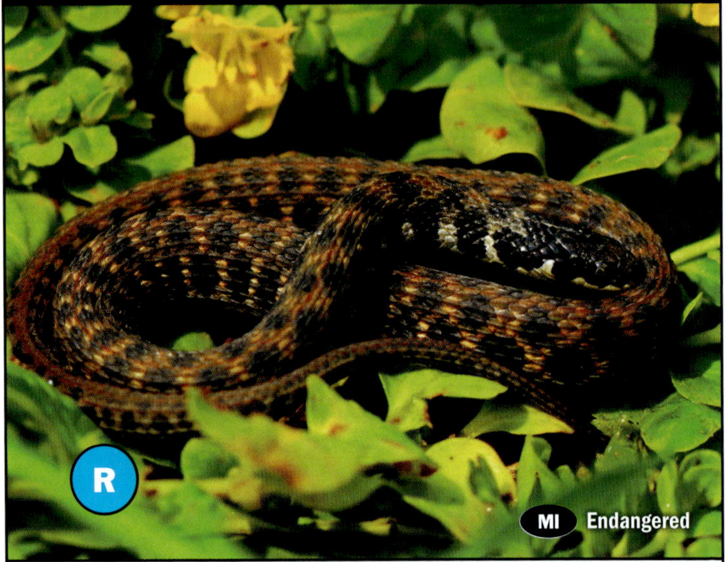

Edges of swamps and marshes, woods, fields, meadows, parks and vacant lots.

Secretive, often hiding under logs, rocks, or sheltering in crayfish burrows. They are seen more often in spring or fall, because in the summer they are nocturnal.

When threatened, it can spread its body flat, and I mean flatter than other snakes, so from above it appears larger. It stiffens its body and remains still, for whatever good that does. If touched it writhes or hides its head under its coils or tries to escape. If captured, it can release a foul smelling musk.

Kirtland's Snake is actually a type of watersnake but is less aquatic than the others. It can swim but is hardly ever found in water.

It has been extirpated from areas, when its vacant lot habitat was developed.

snakes

A small, very secretive snake that can flatten its body to extreme proportions

Description: Length is 14 to 18 inches. Record is 26 inches.

Kirtland's Snake is ***reddish brown or grayish brown with four rows of small dark blotches or spots running down its back and sides.*** These blotches are faded and indistinct in old individuals. Some have a faint dorsal stripe on their back. The scales are keeled. Its ***head is black with white on its mouth and throat.*** Its belly is brick red with a row of round black spots on each side.

Hatchlings are dark without evident spots but with a more intense belly color.

Diet: Mostly earthworms and slugs but also other small invertebrates like crayfish, leeches and water striders.

Juvenile Kirtland's Snakes are darker than adults. The black head and white throat stands out.

Thought rarely found in water, Kirtland's Snakes are closely related to watersnakes.

Courtship, Mating & Young: Mating is in the spring with 4 to 15 young born in late summer or early fall. Like other live birth snakes, the young are enclosed in a membrane from which they soon crawl out. Newborn snakes are ready to defend themselves as they will coil, open their mouths and strike at an enemy. At least some of the females mate in their second spring.

snakes

Range Notes: Not found in Minnesota or Wisconsin. Southwest Michigan to northeast Illinois. Central Illinois and southern Indiana north to southwest Ohio. Scattered in Pennsylvania and Kentucky.

Blue Racer (subspecies of North American Racer)

Coluber constrictor foxii

Dry sunny areas such as hillsides, forest edges, open woods, and grasslands.

Nature Notes:

Blue Racers are diurnal, alert, fast snakes that crawl with the head and neck elevated to better spot the movement of potential prey. They use their speed to catch frogs, lizards, snakes or mice. Bird eggs or young are also on the menu, and the Blue Racer will climb through bushes to get them.

Contrary to their Latin species name, they do not constrict prey but simply overpower it.

The Blue Racer will use speed to escape enemies, but when cornered it will coil defensively, vibrate its tail and bite.

Ten other subspecies are recognized in North America.

snakes

A fast snake that cruises prairies and open hillsides with its head held high as it searches for prey.

Description: Length is 36 to 60 inches. Record is 75 inches.

The Blue Racer is the subspecies found in our area. The ***scales are smooth, and its color is bluish gray***. It has ***large eyes and a white or yellowish chin and throat***.

The belly is light gray, pale blue or yellow.

Juvenile Blue Racers look nothing like the adults. The juveniles are heavily blotched with dark reddish brown on a gray or tan ground color. The head and belly are spotted. Attaining the adult coloration takes two years.

Diet: Frogs, lizards, snakes, mice, bird eggs and young birds. Blue Racers do not kill by constriction.

Note the adult's distinctive yellow throat.

Juveniles are blotched with dark spots and look nothing like the adults (bottom left). This stage lasts a year or two. Adults show either a whitish or yellow chin and throat (top left). Photo on right shows a half-grown Blue Racer; not checkered, not yet blue.

Courtship, Mating & Eggs: The male Blue Racer will follow the pheromone trail of a female and catch up to her. His body ripples with movement as they entwine and mate. In late June or July, she lays 3 to 32, usually 8 to 20 elongated eggs with a granular texture, like they've been sprinkled with salt. They do not adhere to each other and hatch in 43 to 65 days.

Males take one or two years to mature while females require two or three.

Recently emerged juvenile and eggs.

snakes

Range Notes: The Blue Racer subspecies is found in most of southeastern Minnesota and most of southwestern Wisconsin as it seems to have followed the Mississippi, Minnesota, and Wisconsin Rivers. It also inhabits the western and southern areas of Lower Michigan. Illinois, Indiana, northwest half of Ohio.

Prairie Ring-necked Snake (subspecies of Ring-necked Snake) *Diadophis punctatus arnyi*

Bluff prairies and southwestern facing wooded bluffs with lots of rocks.

Nature Notes:

Ring-necked Snakes are very secretive and very rarely bask out in the open. They spend most of their time under rocks or in burrows, as they hide or regulate their body temperature. Over the years, I've found about 200 Prairie Ring-necked Snakes and only three were not underneath a rock.

In the spring and fall, numerous individuals may be found on certain bluffs, as they seem to live in colonies.

Eggs are elongate and become distorted.

This small snake has a unique defensive behavior of coiling its tail and displaying the bright colored underside.

Description: Length is 10 to 15 inches. Species record is 27 11/16 inches.

The Prairie Ring-necked Snake is one of the subspecies in our three state area. It is dark bluish gray with a ***distinctive bright orange collar***. The scales are smooth. The ***underside is bright yellow, changing to reddish-orange as it nears the tail***. The ventral surface is covered with scattered black spots.

Hatchlings have the same colors and patterns as adults.

Diet: Earthworms, slugs, insects, spiders, small salamanders, frogs, lizards and snakes. Ring-necked Snakes are constrictors when necessary. Also, they have enlarged upper rear teeth and toxic saliva which may help subdue their prey.

Juveniles, even recently hatched, match the color of the adults.

Orange belly transitions to red-orange at the tail (bottom left). When threatened the Prairie Ring-necked Snake corkscrews its tail revealing the bright orange underside and protects its head under a coil (right)

Courtship, Mating & Eggs: Breeding takes place soon after emergence from hibernation. The female lays 2 to 10 eggs inside or under rotting logs or stumps or underneath rocks or in burrows. The eggs are long and are somewhat shriveled. After absorbing water, they may appear distorted. They will hatch in 40 to 60 days. There is a single report of a female giving birth to six live young.

Nature Notes:

As a defense mechanism, the Ring-necked Snake will coil its tail into a corkscrew shape and display the bright reddish-orange bottom at the intruder, while hiding its head under a coil. A potential predator may strike at the tail and give the snake a chance to escape.

Range Notes: In southeastern Minnesota and southwestern Wisconsin along the Mississippi and lower Wisconsin Rivers. Ranges south to northwest Arkansas, eastern New Mexico and southeast South Dakota.

Northern Ring-necked Snake (subspecies of Ring-necked Snake) *Diadophis punctatus edwardsii*

Moist, shaded forests or forest edges but not river bottom forests.

Nature Notes:

Ring-necked Snakes are rarely seen because they are secretive, hiding under logs or rocks by day. They come out at night to forage for small prey.

Eastern Red-backed Salamanders are a frequent food item.

Ring-necked snakes are usually colonial, with many inhabiting a small local area.

Unlike the prairie subspecies, Northern Ring-necked Snakes do not coil their tail in a corkscrew shape and hold it up defensively.

A secretive snake not often seen as it hides under rocks and logs.

Description: Length is 10 to 15 inches. Species record is 27 11/16 inches.

The Northern Ring-necked Snake is one of the subspecies in our three state area. It is ***bluish black or bluish gray or brownish, with a bright orange or yellow collar.*** The scales are smooth and shiny. The ***belly is yellow or orange-yellow, and the underside of the tail is not reddish-orange like it is on the Prairie Ring-necked Snake.*** It also has ***fewer black spots*** than the Prairie Ring-necked Snake and sometimes none at all.

Hatchlings have similar colors and patterns as the adults.

Diet: Earthworms, slugs, insects, salamanders, frogs and small snakes. Ring-necked Snakes can constrict their prey when necessary, and

A close up showing the distinctive ring around the neck.

The Northern Ring-neck does not coil its tail like the prairie subspecies.

The belly is bright yellow with black spots.

they have enlarged upper rear teeth along with toxic saliva to help subdue their prey.

Courtship, Mating & Eggs: Mating can take place from spring through fall. The female can store sperm until spring, if that's when she lays the eggs. The one to ten eggs are noticeably elongated and frequently have a distorted shape and tapered ends. The eggs may be placed in the same moist, hidden site selected by other females. Eggs need to incubate 40 to 60 days and they hatch in the second half of August or early September. Maturity isn't attained until the second or third year.

Range Notes: Northeastern Minnesota, the northeast half of Wisconsin, Upper Michigan and western Lower Michigan. Range extends Ontario to Labrador and southwest to northwest Alabama.

Plains Hog-nosed Snake

Heterodon nasicus

Open sandy prairies, river terraces with sand dunes or gravel.

Nature Notes:

Toads are a specialty, and this snake is well equipped. Its turned up nose is for burrowing after them. It is immune to their toxic skin secretions. When the toad inflates its body, hoping to be too big to swallow, the snake punctures it with its enlarged rear teeth.

The Plains Hog-nosed Snake usually avoids detection with its cryptic coloration. That failing, it tries a bluffing threat display. It spreads its neck out like a cobra and hisses loudly with its mouth gaped wide. It may strike with its mouth closed. Then it turns upside down and lays still. Looks dead. Smells dead. After a short while it starts flicking its tongue, turns upright and crawls away.

A stocky little snake with a ***sharply turned up nose*** it uses for digging into the sandy soil.

Description: Length is 15 to 25 inches. Record is 36 1/8 inches.

Formerly called the Western Hog-nosed Snake, the Plains Hog-nosed Snake has a ***light brown body with dark brown blotches down its back*** and alternating smaller dark brown blotches down its sides. It's a stout snake with keeled scales, including its upturned rostral scale.

The ventral plates of the Plains Hog-nosed Snake are jet black with white and yellow squares.

Hatchlings have similar colors and patterns to the adults.

Similar Species: *The upturned nose is more apparent than it is on the Eastern Hog-nosed Snake* (see inset photo above).

snakes

Toads are a primary food. The toad inflates itself hoping not to be swallowed, but soon one of the snake's enlarged rear teeth will puncture it.

Pattern and coloration provide excellent camouflage for their gravelly sand prairie habitat.

Diet: Toads, of course, but also frogs, salamanders, skinks, other snakes, mice, ground-nesting bird eggs and turtle eggs. They have been known to follow the scent trail of a female turtle to her nest.

Courtship, Mating & Eggs: Plains Hog-nosed Snakes will mate in fall and spring. Males will search for and follow scent trails left by females, eventually catching up and mating with them. About 30 days later, the female will lay 2 to 24 eggs in the sandy soil. They will hatch in 47 to 75 days.

Range Notes: Broad range from Saskatchewan to Texas. Isolated in northwest and eastern Iowa and northeast and central Illinois.

Hatching from its thin-shelled egg

Eastern Hog-nosed Snake

Heterodon platirhinos

Sandy soil habitats such as sand prairies, oak savannas, and open pine or deciduous woods.

Nature Notes:

The Eastern Hog-nosed Snake flattens its body and spreads out its neck like a cobra. It coils and gapes its mouth while hissing loudly. It may strike with its mouth closed. If bluffing fails, and the hognose is gently poked a couple times, it starts writhing and twisting. It may empty its cloaca of stinking wastes and musk. It rolls onto its back and lies still with its tongue hanging out. It looks and smells dead. If turned over, it writhes around, turns on its back, and again lies still. It's like the snake knows that the only dead snake is an upside down snake. If left alone for a few minutes, the hog-nose lifts its head and starts flicking its tongue. It rolls upright and crawls away.

A toad's worst nightmare, this snake has numerous adaptations to overcome the toad's defenses.

Description: Length is 20 to 33 inches. Record is 49 15/16 inches.

The Eastern Hog-nosed Snake has a specialized upwardly curved rostral scale, forming its characteristic ***upturned nose.*** This stocky snake has keeled scales, including the rostral scale (the rostral is both upcurved and keeled). Most have large blotches down the back alternating with smaller ones on the sides. Colors are variable with yellow, black, gray, brown, reddish brown, tan, olive, and other colors possible, most of which can be found on a single snake. ***Most have a pair of large black spots behind the head, which resemble large eyes when the hog-nose flattens its neck defensively.***

Juvenile Eastern Hog-nosed Snakes are colored and patterned like adults.

Toads are a favorite food. Mildly toxic saliva may help subdue prey.

Hog-nosed Snakes are great actors; they play dead to deter predators (right middle and bottom). But put him back on his belly and he will writhe around and flip back over! The "cobra act" displays black "eyes" (top right).

Similar Species: (see Plains Hog-nosed Snake).

Diet: That upturned nose is for digging after toads. The toad's skin toxins don't deter hog-nosed snakes. The toad inflates its body to avoid being swallowed whole, but a pair of enlarged upper rear teeth puncture the doomed toad. Mildly toxic saliva may help subdue prey. Also eats frogs, salamanders, small reptiles, birds and small mammals.

Hatchlings can bluff and play dead just like the adults. Egg shells are thin.

Courtship, Mating & Eggs: Lays 4 to 61 eggs, which are thinner than most snake eggs. They hatch in 50 to 65 days. Hatchlings can immediately perform their bluffing and death acts.

Range Notes: From our region east into Ontario, Massachusetts and south to Florida and east Texas and scattered in Nebraska.

Eastern Milksnake (subspecies of Milksnake)

Lampropeltis triangulum triangulum

Rocky hillsides, in bottomland woods, open woods, or fields. Cover, such as flat rocks, is important.

Nature Notes:

Milksnakes are more common than most people realize because of their secretive habits. They hide under flat rocks, boards, tin, or in burrows. Milksnakes are mostly nocturnal, especially in the summer. When they are out during daylight hours, they have effective cryptic coloration.

Milksnakes often share a hibernaculum in the same southwest facing bluff localities or hillsides with Western or Gray Ratsnakes and Timber Rattlesnakes.

When threatened, Milksnakes will coil and vibrate their tails, producing a buzzing noise in dry leaves or brush.

snakes

A supreme mouse-eating snake that constricts its prey to death before devouring it.

Description: Length is 24 to 36 inches. Record is 52 inches.

The Eastern Milksnake is the subspecies found in our three states. The ***head is reddish brown or brown with usually a light Y- or V-shaped blotch on the nape***. It has smooth scales with ***large saddle-like reddish brown dorsal blotches.*** Alternating with them are smaller blotches on the sides. The ground color is gray or tan. The belly is white, checkered with black.

Juveniles are light gray with bright red blotches.

Diet: Milksnakes are constrictors, feeding on mice and voles, which they grab in their jaws, wrap in a couple coils and suffocate them. They also eat snakes, lizards, birds and bird eggs. Juveniles, especially, feed on other juvenile snakes.

The Milksnake's round pupils and lack of a facial pit indicates that it's a harmless snake.

Cryptic coloration helps make Milksnakes virtually disappear on lichen-covered rocks.

Juveniles have bright red blotches bordered by black.

Courtship, Mating & Eggs: Milksnakes mate when they emerge from hibernation or later in the spring. The male follows a female's pheromone trail until he finds her and presses his chin down on her neck. Sometimes he holds her neck in his mouth, and they intertwine. In late June or early July, the female lays 3 to 24 eggs, which usually stick together in a clump. Eggs sticking in a clump means they can't be eaten by another snake. The female lays them in a moist place like the inside of a rotten log. After 6 to 9 weeks the eggs hatch. Milksnakes mature in their third or fourth year.

Egg is slit by young's egg tooth. Milksnake eggs adhere together in a clump.

Range Notes: From our region to southern Ontario to southern Maine and southwest to northeast Louisiana and northwest to northeast Oklahoma and western Iowa.

Plain-bellied Watersnake

Nerodia erythrogaster

Shallow ponds, swamps, sloughs and slow-moving streams and connecting upland corridors to permanent water.

Nature Notes:

Plain-bellied Watersnakes are more likely than other watersnakes to be encountered on land, as they move from small drying ponds to a larger body of water. Some have been known to actually leave the water and try to escape on land when threatened.

Defensively, they flatten their bodies and strike repeatedly and emit an odorous musk. If close to water, they can quickly dive to the bottom and hide under submerged rocks or logs for up to an hour.

In the fall, they seek refuge in burrows or deep crevices in rock formations on higher ground. Some choose crayfish burrows but risk drowning if the water level goes up while they are dormant.

snakes

Primarily a snake of the southeast U.S but an isolated population extends into Michigan.

Description: Length is 30 to 48 inches. Record is 64 3/8 inches.

The Plain-bellied Watersnake is a plain-bodied watersnake with a solid dark color: black or dark gray or dark brown keeled scales. The only bright color usually seen is the ***orange or reddish labials (lips) and its throat***. Sometimes there is a trace of darker juvenile patterning. The ***belly is plain, unmarked and light orange, reddish or a coppery color***.

Juveniles look totally different with a strongly blotched pattern of dark brown on reddish brown. Large blotches run down the back and alternate with smaller blotches on the sides.

Diet: Frogs and tadpoles, salamanders and their larvae, fish and crayfish. Sometimes they trap fish and tadpoles with their bodies in shallow water while they feed on them.

Belly color can range from coppery to reddish to light orange.

Courtship, Mating & Young: Plain-bellied Watersnake mating activity peaks during May, shortly after emergence from hibernation, and the snakes are still gathered near spring basking sites. Any time from mid-September to mid-October, a mated female will have from 2 to 55 young, measuring 8 to 12 inches long.

Range Notes: In our region restricted to an isolated population in southern lower Michigan. Southeast Kansas and southern Illinois to Gulf of Mexico and east to North Carolina.

There were formerly several subspecies but they are now all lumped under *Nerodia erythrogaster*.

snakes

Unlike adults, juveniles are heavily patterned.

Northern Watersnake (subspecies of Common Watersnake) *Nerodia sipedon sipedon*

Permanent bodies of water with logs or open banks for basking–rivers, sloughs, lakes, marshes, bogs, ponds.

Nature Notes:

Although watersnakes eat fish, they have no impact on game fish.

Watersnakes have nasty dispositions and will bite viciously.

Some people mistake them for the cottonmouth, but the cottonmouth has a thick, blocky looking head and vertical pupils. Also, cottonmouths get nowhere near Minnesota, Wisconsin, or Michigan.

Watersnakes are often seen basking on logs, rocks or river banks, where they are usually alert and quick to escape into the water. They swim with the head out of the water, or they dive and find a place to hide while submerged. They can stay under water for at least an hour.

A heavy-bodied snake that has earned a reputation for having a nasty disposition.

Description: Length is 22 to 42 inches. Record is 59 inches.

The Northern Watersnake is the subspecies found in our area. The scales are keeled. The larger snakes are ***heavy-bodied***. They can be ***gray, tan, or brown with dark brown or reddish brown bands on the front of the body. The bands are wider than the areas between them.*** Farther back the bands break up and become disconnected. The pattern is most vivid on younger snakes, often fading and darkening with age. ***Older, larger snakes often have little visible pattern, especially if dry***. The belly is white with red or brown half moons.

Juveniles have the same color and pattern as adults.

Diet: Mostly cold-blooded vertebrates, such

A large, heavy-bodied Northern Watersnake basks in the quiet backwater or a river.

A juvenile watersnake eats a minnow (right). Fish is a staple for this water-loving snake. The patterns and colors are most distinct and brightest on younger individuals. Wet adults can also look more patterned and colorful.

as fish, frogs, tadpoles, or salamanders. Prey is hunted on land or in the water, and smaller prey can be swallowed while submerged.

A rare photo of a wild Northern Watersnake hunting fish or frogs in a fastwater stream.

Courtship, Mating & Young: Northern Watersnakes mate on land or in the water. The male moves up beside a female or over her and presses his chin down on her neck and back, while he lines up his hemipenis with her cloaca. After a gestation period of two months, females give birth to 4 to 99 young, most often 15 to 40. It takes two or three years before the young reach maturity.

Range Notes: From our region east to Maine, southwest to Ohio, south to northwest South Carolina.

Smooth Greensnake

Opheodrys vernalis

Moist grassy areas associated with lakes, marshes, bogs, and meadows. Also, open woods and oak savannas.

Nature Notes:

Smooth Green Snakes are well camouflaged as they move through grass and the leaves of low bushes. When not foraging, they are usually hiding underneath something, such as a flat rock or log.

Unfortunately, these harmless, beneficial snakes have been adversely affected by the widespread use of pesticides. If it doesn't kill the snake directly, it poisons their food.

Sometimes hibernate together with Red-bellied Snakes and Plains Gartersnakes in abandoned ant mounds.

Death causes the unstable yellow pigment to break down and the greensnake turns blue, but a similar sized Blue Racer would still have juvenile spotting.

A small, beautiful leaf green snake that feeds almost exclusively on insects.

Description: Length is 12 to 20 inches. Record is 26 inches.

The Smooth Greensnake has ***smooth scales that are leaf green*** and identification is unmistakable except when it's a different color. Instead of green, some are actually tan or rarely, bronze; especially in northern and central Wisconsin. ***The labials (lips), throat and belly are white or pale yellow.***

Hatchlings are 4 to 5 inches long and dark olive green, no matter what color phase they will be as adults.

Diet: Insects, such as crickets, grasshoppers, beetles, and hairless caterpillars. Also, spiders and small salamanders.

Courtship, Mating & Eggs: Smooth Greensnakes usually breed in spring, but the

Eggs are laid in a moist hidden spot such as, in this case, the inside of a rotting log.

All young are dark when they first emerge from the egg.

This young snake is just starting to turn green.

A rare threatening pose from a harmless adult snake.

eggs are retained longer inside the female and egg-laying may extend into September. The female selects a moist, hidden site, such as inside or underneath a rotting log and lays from 2 to 13 eggs. Sometimes more than one female chooses the same nesting location. The well-developed eggs hatch in only 4 to 30 days and have the shortest incubation time of any snake species in the United States. The young probably reach breeding size in their second year.

Difficult to see in the grass, but not hard to spot when crossing a gravel road!

Range Notes: Throughout Minnesota except the northeast and extreme southwest, most of Wisconsin except the southwest and almost all of Michigan. From our region northwest to Saskatchewan south to Utah, east to Labrador, and south to Virginia.

Western Ratsnake

Pantherophis obsoletus

The north and east side of bluffs and hills. Old abandoned houses, barns or sheds are also frequented.

Nature Notes:

Western Ratsnakes are arboreal, using their flat bellies with squared edges, to climb straight up tree trunks and may be 30 to 40 feet high. Forty years ago, I saw photos of a Western Ratsnake high in a tree constricting a squirrel. I still haven't seen that in real life.

In the fall, Western Ratsnakes will move to the southwest side of bluffs and hibernate with Timber Rattlesnakes and Blue Racers.

snakes

An arboreal, squirrel-eating snake that barely gets into Minnesota at the extreme northeast edge of its range.

Description: Length is 42 to 62 inches. Record is 86 inches.

Formerly, this species was called the Black Ratsnake and it included snakes on both sides of the Mississippi River, until it was determined by morphology and mitochondrial DNA that they were two different species, divided by the Mississippi. East of the Mississippi, it's called the Gray Ratsnake, *Pantherophis spiloides*. ***This is the species found on the west side of the river in Minnesota.***

The Western Ratsnake in Minnesota has a ***white chin and throat*** and its body is blotched. The blotches might be dominant or not so evident, if the body is mostly black. Between the scales are flecks of red or white. The scales are

The Western Ratsnake has white labials and a white throat.

This one was found crawling on the forest floor, likely in search of mammalian prey.

weakly keeled. Its belly is flat with squared corners on the edges.

Hatchlings are similar to hatchling foxsnakes, but the former's top rear of the head is plain with no bold black markings.

Diet: Mammals such as squirrels, chipmunks, mice and voles. Occasional birds and their eggs. Mammalian prey is constricted and suffocated along with adult birds.

Courtship, Mating & Eggs: The female releases a pheromone to attract males and they can follow her trail. The male approaches the female with a jerking motion, pressing his chin down on her neck. Rippling movements travel up his body as he aligns it with hers. She lifts her tail and they copulate for about 20 minutes. She will lay 6 to 30 eggs inside or under rotted logs or stumps or in piles of sawdust. They adhere to each other and hatch in 60 to 75 days.

Range Notes: Only found west of the Mississippi River. Extreme southeastern Minnesota south to Gulf of Mexico and west to Kansas and central Texas.

Gray Ratsnake

Pantherophis spiloides

The north and east sides of wooded hills and bluffs. Shrubby fields with abandoned houses, barns or sheds.

Nature Notes:

Gray Ratsnakes are arboreal, using their flat ventral scales with squared edges, they can climb straight up the trunk of a tree and may be found 30 to 40 feet high.

Gray Ratsnakes will grab and constrict their mammalian prey to death before swallowing it whole.

In the fall they move to the south and west sides of hills and share hibernacula with other snakes.

snakes

An arboreal, squirrel-eating snake that is found in southwestern Wisconsin.

Description: Length is 42 to 72 inches. Record is 84 ¼ inches.

Formerly this snake was called the Black Ratsnake until it was found (via morphology and mitochondrial DNA) that there were actually two species divided by the Mississippi River. Now ***this species is the snake found in Wisconsin and Michigan.*** And the snake in Minnesota is called the Western Ratsnake, *Pantherophis obsoletus.*

Gray Ratsnakes in our area have a ***white chin and throat*** with a blotched body that is mostly black. In between the weakly keeled scales are flecks of red, orange, yellow or white. The belly is flat with square corners at the edges.

Hatchlings look remarkably like hatchling foxsnakes but lack the heavy markings on the top

If you eat squirrels, you better be able to climb trees.

Ratsnakes can use their flat belly scales to climb straight up trees for food and security.

rear of the head. Their appearance will start to change after two years.

Diet: Gray squirrels, red squirrels, chipmunks, voles and mice; birds and bird eggs.

Courtship, Mating & Eggs: Preceding courting, two males may battle for supremacy by trying to raise up higher than the other and pushing the other male down. Finally, the loser will crawl away. The male is able to follow a female's scent and approaches her with jerking movements. He presses his chin down on her back as rippling waves move up his body, and he aligns his body with hers. They copulate for about 20 minutes. In late June or July, the female will lay 5 to 44 eggs inside or under a rotten stump or log or in a sawdust pile. Sixty to 75 days later they will hatch.

Range Notes: Southwestern Wisconsin and southern Michigan. South to southeast Louisiana and northeast to Ontario, Quebec and southern Maine.

Western Foxsnake

Pantherophis ramspotti

Forest edges, bottomland forest, pine barrens, old fields, marshes, and railroad beds through wetlands. Usually near water.

Nature Notes:

Some people think the foxsnake must be venomous because of its vibrating tail, blotched body and copper colored head. Many snakes have been killed because of this ignorance. Fortunately, many escape notice because their coloration and pattern provides excellent camouflage.

Hibernate by migrating to a hillside hibernaculum or a nearby slope or old foundation. Surprisingly, they also hibernate submerged in old wells.

Foxsnakes have a musky odor similar to a Red Fox.

snakes

A four to five foot mouse-eating snake that is often found on railroad beds cutting through wetlands.

Description: Length is 24 to 54 inches long. Record is 61 inches.

The scales are keeled dorsally but are smooth on the sides. The ***adult's head is orange-brown, brown, or grayish brown. An average of 37 large brown saddle-like blotches run down the back***, with alternating smaller blotches on the sides. The ground color is yellowish brown or grayish brown. The belly is yellowish with small dark squares, and it is flat with squared edges. ***Found west of the Mississippi River.***

Juveniles are gray with dark blotches. The head is marked with a dark line connecting the eyes, and another line extends from each eye to the angle of the jaw. Distinct black markings are also on top of the head.

Foxsnakes often hunt along railroad right-of-ways.

The dark mark on the back of the head (left photo) distinguishes it from the juvenile Western Ratsnake.

Diet: Kills mice and voles by constriction. Squared ventral scutes help them climb into trees for birds and bird eggs. Juveniles eat worms, frogs and small reptiles.

These eggs were just laid and will hatch in two months.

Courtship, Mating & Eggs: Males will combat each other for dominance by trying to raise up high enough to push over and pin his opponent to the ground. A male moving beside a female will adopt a jerking motion. She might move in the same way. The male takes hold of her neck or head with his mouth. They intertwine bodies and mate.

By the middle of July, the female lays 7 to 27 eggs which adhere to each other. The young hatch in late August or early September.

Range Notes: Recent research has restricted the Western Foxsnake to only west of the Mississippi and St. Croix Rivers. Also Iowa and into Nebraska.

snakes

Eastern Foxsnake

Pantherophis vulpinus

Forest edges, bottomland forest, pine barrens, old fields, marshes, railroad beds through wetlands. Usually near water.

Nature Notes:

Foxsnakes get their common name from their musky odor which is similar to a Red Fox.

Some people think a foxsnake must be venomous because of its vibrating tail, blotched body, and copper colored head. Many snakes are killed. Fortunately, many escape notice because their color and pattern provides excellent camouflage.

Hibernate by migrating to a hillside hibernaculum, old foundation or the end of an old railroad bed. Surprisingly, they also hibernate submerged in old wells.

snakes

A four to five foot, brown-blotched snake buzzing its tail as a warning like a rattler.

Description: Length is 24 to 54 inches. Record is 67 1/8 inches.

The dorsal scales are weakly keeled and the side scales are smooth. The ***brown or orange-brown head is unmarked. An average of 43 dark brown saddle-like blotches run down the back,*** with alternating small side blotches between them. The ground color is light brown or yellowish brown. The belly is flat with squared edges and is yellow with black squares. ***Found east of the Mississippi River.***

Hatchlings are light gray with dark blotches. The head has a black line running between the eyes and a black line between the eye and corner of the mouth.

Diet: Kills mice and voles by constriction. Squared ventral scutes help them climb trees

It's not unusual to find a Foxsnake swimming, but you're more likely to find one crossing a road.

This small Foxsnake is about a year old.

An adult Foxsnake doing his best threat display, but he never tried to bite.

for birds and bird eggs. Juveniles eat worms, frogs and small reptiles.

Foxsnakes climb trees in search of birds and bird eggs, but this one has followed a female to a hole in a tree.

Courtship, Mating & Eggs: Males combat for dominance by raising up high enough to push his opponent over and pin him to the ground.

A male can follow a female's scent and crawls with a jerking motion when he is beside her. She may start jerking also. He grips her head or neck in his mouth, they intertwine bodies and mate.

In July, the female lay 7 to 27 eggs which adhere to each other. They hatch in about 60 days.

Range Notes: Recent research has expanded this snake's range west to the Mississippi and St. Croix rivers, which is now the boundary between Eastern and Western Foxsnakes. Also northern Illinois and northwest Indiana.

Bullsnake (subspecies of Gopher Snake)

Pituophis catenifer sayi

Sand prairies, bluff prairies, oak savannahs, pastures and other open areas with sandy soil.

Nature Notes:

Bullsnakes are impressive with their defensive behavior: coiling, vibrating the tail, striking and the loudest hiss of any of our snakes. But their first response would be to quickly disappear down a nearby burrow.

Bullsnakes are constrictors and suffocate their mammal prey in their muscular coils or press them against the wall of their own burrow before devouring them.

Found mainly in open sandy landscapes.

A big, constricting prairie snake that feeds on ground squirrels and gophers.

Description: Length is 37 to 72 inches. Record is 105 inches for the species.

The Bullsnake is the subspecies in our three state area and is ***our largest snake***. Its relatively small pointed head is unique with an ***enlarged rostral (nose) scale***. The ***head is yellow, straw yellow, or brown with black markings and the upper labials (lip) have black bars***. It is blotched with the blotches usually black on the neck and anterior part of the body and changing to brown blotches on the middle of the body and then to black again. ***Nearing the tail, the colors are black on straw yellow and continues to black rings on the yellowish tail***. It has keeled scales.

Newly hatched, the snakes are grayish with dark blotches but have the same colors and

View of head with reinforced rostral.

Just hatched from its leathery shell.

This juvenile was found crossing a road.

This small Bullsnake was found in a hollow in a stump.

pattern as adults after their first shed.

Diet: Pocket gophers, ground squirrels, voles, mice, rabbits and ground nesting birds.

Courtship, Mating & Eggs: The female attracts males to her by giving off pheromones, and they can then track her. A male will crawl with a jerking motion and grasp her head, neck or back in his jaws. When she is ready, she will lift her tail and the male will insert his hemipenis. In late June or early July, the female will lay 3 to 24 eggs into an old mammal burrow or a space she excavated under the sandy soil. The leathery eggs adhere to each other. They hatch in about two months.

Range Notes: From our area to extreme northwest Indiana, central and northwest Illinois. Eastern edge and across central and southwest Iowa. Western Missouri and northwest to Alberta and south to Texas and Mexico. Seven other subspecies are found to the West Coast.

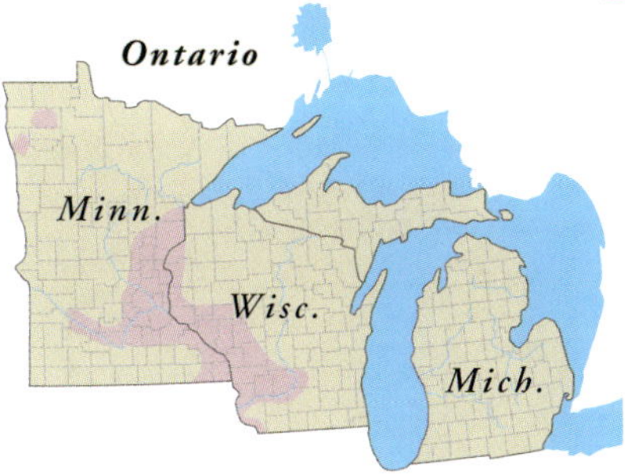

Queensnake

Regina septemvittata

Small, clear, spring-fed creeks or streams with rocky bottoms and sunlit places for basking.

Nature Notes:

Queensnakes are more secretive than watersnakes. They rarely bask, usually in bushes that overhang their stream, and are quick to dive into the water if disturbed. There they often hide under submerged rocks or just swim farther away from the intruder. The vast majority of Queensnakes that have been found were underneath flat rocks that were submerged, partially submerged or on the shore.

Queensnakes retreat below the frost line into crayfish burrows, mammal burrows or inside deep rock crevasses to survive the winter.

This snake feeds almost exclusively on newly molted crayfish.

Description: Length is 15 to 24 inches. Record is 36¼ inches.

The Queensnake has a *small head, hardly bigger than the neck.* Its *rostral scale (nose) is light yellow or cream colored, as are the labial scales (lips)*, and this *light yellow or cream color extends as a stripe all the way down the side of the body.* Above the stripe, the body is brown, gray, or olive. Three black stripes, one dorsal and one on each side, are sometimes visible. The scales are keeled. The *belly is light yellow or cream colored with four brown stripes.*

Juveniles are miniature versions of the adults but with a more distinct pattern.

Diet: Mostly crayfish that have recently shed the hard exoskeleton. After shedding, a cray-

The Queensnake's belly shows four stripes.

Close-up of the white labial scales and the white side stripe.

fish will hide until its new exoskeleton hardens, but molting crayfish release a hormone which can be detected by the snake. The Queensnake searches them out with its forked tongue and Jacobson's Organ.

Occasionally frogs, tadpoles, newts or minnows are taken.

Courtship, Mating & Young: Queensnakes mate in the spring but may have mated in the fall as well, and the female bears 5 to 39 young in late summer. Males are sexually mature in two years, and females probably breed in their third year.

Range Notes: From our area they range to northeastern Illinois, central Indiana, Ohio and western Pennsylvania. Scattered in southern Ontario. southeast Pennsylvania to eastern Mississippi.

DeKay's Brownsnake

Storeria dekayi

Bogs, swamps, marshes, moist woods and wooded hillsides.

Nature Notes:

DeKay's Brownsnakes are most often seen on country roads during their fall migration from moist bottomlands to wooded hillsides where they will hibernate. They will do a reverse migration in the spring to get to their summer habitat. They are diurnal during these spring and fall migrations.

In the summer, when the weather is warmer, they are normally secretive, hiding under rocks or logs by day and coming out at night to feed.

DeKay's Brownsnakes often hibernate with Red-bellied Snakes and Smooth Greensnakes in burrows, old foundations, rock piles, and ant mounds.

snakes

A small snake that can appear in large numbers during spring or fall migrations.

Description: Length is 9 to 13 inches. Record is 20 ¾ inches.

DeKay's Brownsnake, formerly divided into subspecies, is brown, dark brown, reddish brown, or grayish brown. The ***middle of the back is usually lighter in color and bordered by small dark spots.*** Some spots opposite each other are linked with a dark line. The belly is buff, cream-colored or pinkish. The scales are keeled.

Juvenile DeKay's Brownsnakes are similarly colored but have light patches behind the head, which are sometimes connected, forming a collar.

Diet: Slugs, worms, soft-bodied insects and insect larvae. They also feed on snails by employing specialized jaws and teeth to pull the soft body from the shell.

Every spring brownsnakes migrate from hibernacula to their summer habitat.

Brownsnakes bear live young.

Courtship, Mating & Eggs: In the early spring, before dispersing from their hibernacula to summer habitat, a male will locate a female by following her pheromone trail. He follows her scent with his flicking tongue and moves beside her. He presses his chin on her neck and may have rippling movements going up his body. He inserts a hemipenis into her cloaca.

She gives birth to 3 to 41 young, usually about a dozen, and usually in late July or August. There is no maternal care, although they may be found together for a few days.

Range Notes: From our area range extends east into southeast Maine, south into the central Florida panhandle, west into central Texas and north into the southeast half of Iowa.

snakes

Red-bellied Snake

Storeria occipitomaculata

Boreal forests, hardwood forests, open woods, sphagnum bogs, and marshes.

Nature Notes:

Red-bellied Snakes migrate to hibernacula, such as abandoned ant mounds, where they hibernate with brownsnakes, greensnakes and gartersnakes. In the spring they emerge and do a reverse migration. During these migrations, they often cross roads where they sometimes linger to warm themselves. Their small size on a gravel road is easy to overlook and many have been killed.

Red-bellied Snakes are harmless and hardly ever attempt to bite, but they curl their lip to look fierce and they flatten their body and release musk when threatened. They also have been known to play dead.

A very small snake with gray or brown color phases and a brick red belly.

Description: Length is 8 to 10 inches. Record is 16 5/8 inches.

Red-bellied Snakes have two basic color phases: gray or reddish brown. Rarely, a black one is found. They may have a lighter dorsum like the Brownsnake, but they lack the rows of spots and usually have two or four thin black longitudinal stripes. The scales are keeled. The ***belly is bright red***; usually. Red-bellied Snakes have been found with ventral colors of orange, light yellow, pink, or even gray or bluish black. Red-bellied Snakes have ***three tan or yellowish spots behind the head***: one on the nape and one on each side of the neck. These spots may be merged into a collar.

New-born snakes look similar but may be darker and most have a light colored collar or at least a bright spot on the back of the neck.

This is the reddish-brown phase of the Red-bellied Snake. They can also be dark gray.

The red ventral scales create the red belly it's named for.

Diet: Slugs, earthworms, beetle larvae and other small, soft-bodied creatures. Also, snails, which Red-bellied Snakes feed on by extracting them from their shells with teeth and jaws specialized for that purpose.

Courtship, Mating & Eggs: Red-bellied Snakes probably mate in the spring like most snakes, but they are also known to mate in the summer and in the fall too. Sometime during late July, August, or early September, a mated female will give birth to 1 to 23 young, with seven or eight being an average number.

Range Notes: Range extends barely to Saskatchewan and northeast North Dakota; isolated in Black Hills of South Dakota. South to extreme east Texas. East to northern Florida and north into Ontario, Quebec and Labrador.

Orange-striped Ribbonsnake (subspecies of Western Ribbonsnake) *Thamnophis proximus proximus*

Ponds, streams, rivers, sloughs, marshes and lakes with thick vegetation on the shore or at the edge.

Nature Notes:

Orange-striped Ribbon Snakes are usually diurnal, but in hot weather, they become active at dusk or after dark.

Orange-striped Ribbon Snakes climb into bushes near or overhanging the water, where they can look for prey items. They are alert and quick to escape a land-based threat. They are fast and agile as they flee into the water and dive to safety.

snakes

The orange stripe makes this snake look like a slim version of the Plains Gartersnake.

Description: Length is 20 to 30 inches. Record is 49 7/8 inches.

The Orange-striped Ribbonsnake is the subspecies in our area. Basically, this is ***a slim, long-tailed gartersnake with white labial scales (eight on each side of the upper jaw) and a white chin and throat.*** There is also a ***vertical white scale in front of its large eye.*** A pair of small bright yellow spots (parietal spots) are usually touching on the top of its head. The scales are keeled. There are ***three longitudinal stripes, with the center one usually being bright orange***, and the side stripes, on scale rows three and four, are yellow or greenish white. These contrast with the black or very dark brown upper body color.

Juveniles are colored and patterned the same as the adults.

Notice the white lips and vertical white bar in front of eye.

This is a very slender snake, that gives the appearance of a slim gartersnake.

Diet: Newly metamorphosed frogs and toads, tadpoles, small salamanders, salamander larvae, and minnows comprise most of their diet.

Courtship, Mating & Eggs: Orange-striped Ribbonsnakes mate in the spring and in late summer the females give birth. A female can have 4 to 36 young. Most mature in about two years, but females may not breed until their third year.

Range Notes: Orange-striped Ribbonsnakes are rare in Wisconsin and occur in widely isolated areas in the southern half of the state.

Also, a few scattered sites in Indiana and eastern Illinois plus southeast Iowa south to northeast Texas and central Louisiana.

Northern Ribbonsnake (subspecies of Eastern Ribbonsnake) *Thamnophis saurita septentrionalis*

Sphagnum bogs and the edges of streams and marshes where there is plenty of low vegetation.

Nature Notes:

Northern Ribbonsnakes cruise the surface of shallow water and move through shoreline vegetation as they hunt various amphibians. They often climb into bushes, like blueberry or leatherleaf, to bask in the sun but are wary and quick to escape if disturbed.

Hibernation can be in low places, such as crayfish burrows, or in sites on higher ground with other snakes.

Northern Ribbonsnakes have a defensive strategy not shared by other snakes: Half of their long tail can easily break off, although, unlike lizards, it will not grow back.

snakes

A long slender snake that lives at the edges of bogs where they move easily through thick brush.

Description: Length is 18 to 26 inches. Record is 40 15/16 inches.

The Northern Ribbonsnake is the subspecies found in our three states. The ***head and eyes are large, with a white bar in front of the eye.*** There are seven white upper labials with no vertical dark edges. If there are parietal spots (top of head), they are usually not touching. The scales are keeled. The tail is long, being between one fourth and one third of the total length. The ***three stripes may be yellow, greenish yellow, or white, but the middle stripe could be brown. The lateral stripe is on scale rows 3 and 4 with a brown stripe below.*** The belly is pale green, yellowish or white.

A predator that grabs the Northern Ribbonsnake by the tail may be in for a surprise; their tail can break off allowing the snake to escape. Unlike in lizards, it will not grow back.

Juveniles have the same coloration and pattern as adults.

Diet: Salamanders and their larvae, small frogs and tadpoles, small fish, spiders and caterpillars. Surprisingly, they don't eat earthworms.

Courtship, Mating & Eggs: Northern Ribbonsnakes breed soon after emerging from hibernation, with the male following the female's scent trail. He moves along side her and presses his chin down on her neck. He may have rippling motions going up his body. When she lifts her tail, he inserts his hemipenis.

In late July or August, the female crawls to somewhat higher ground to bear her 3 to 26 young. They will mature in their second or third year.

Range Notes: Throughout Lower Michigan and relict populations occur in Wisconsin.

From our region ranges to northern Indiana and east to Ontario and southeast Maine.

Butler's Gartersnake

Thamnophis butleri

Grassy areas adjacent to marshes and other wetlands, prairies, fields, and vacant lots.

Nature Notes:

Darting away into tall grass or brush is the normal gartersnake escape strategy. Butler's Gartersnake is no exception unless protective cover is too distant. The snake then whips its head back and forth while looping its body from side to side. This bizarre behavior is unique. The exaggerated motion may somehow confuse a predator.

They can emit a smelly musky odor, but it's less foul than that of the Eastern Gartersnake.

Docile and hardly ever bites when picked up.

snakes

This snake, if caught out in the open, goes into a bizarre back and forth whipping motion.

Description: Length is 15 to 20 inches. Record is 29 inches.

Butler's Gartersnake has keeled scales and a small head, only slightly larger than the neck. There are ***three yellow or orange stripes, one dorsal and one on each side. The side stripe is located on scale row 3 and adjacent halves of rows 2 and 4*** (on anterior of body). The ground color is brown, black or olive, and sometimes contains two rows of black spots. The belly is greenish and edged with brown.

Juveniles have the same colors and pattern as adults.

Diet: Butler's Gartersnake feeds mostly on earthworms, but small amphibians are also prey items.

Note that the Butler's Gartersnake has a small head compared to the body.

The side stripe is on scale row 3 and parts of rows 2 and 4.

Courtship, Mating & Eggs: Butler's Gartersnakes breed after emerging from hibernation. Many snakes may be at the site and breeding can be very competitive. Receptive females are identified by scent and one or several males will attempt copulation. A successful male leaves a mucus-like plug in the female's cloaca, which prevents other males from breeding with her for several days. Females give birth to 4 to 20 young, usually in late July or August.

Nature Notes:

Because these snakes often live in vacant lots in urban areas, populations are sometimes destroyed when the lot is developed.

Butler's Gartersnakes hibernate locally in rodent or crayfish burrows or even in large ant mounds.

snakes

Range Notes: Inhabits most of eastern Lower Michigan plus a small area of the southwest corner. In Wisconsin it's found in the Milwaukee and Racine areas.

Ranges to the southern tip of Ontario, scattered in northern Indiana and northern and central Ohio.

Plains Gartersnake

Thamnophis radix

Prairies and moist grassy areas, especially near open rivers, ponds, lakes and marshes.

Nature Notes:

Formerly the Plains Gartersnake was divided into two subspecies.

Hibernation sites include abandoned ant mounds with Smooth Greensnakes and Red-bellied Snakes, old foundations and burrows.

Defensively, if they can't escape, they bite and release foul smelling musk. They also might hide the head under coils and raise the tail, waving it from side to side.

A snake of open areas with a bright orange stripe down its back.

Description: Length is 15 to 28 inches. Record is 43 inches.

The Plains Gartersnake has ***three longitudinal stripes on a ground color of dark brown, olive or black***. The ***dorsal stripe is orange or yellow-orange and the two side stripes, on scale rows 3 and 4, are lighter yellow-orange or yellow***. Unless the ground color is black, there may be two rows of alternating black squarish spots between the stripes. Underneath the side stripe is another row of black squarish spots.

The body is stout and the scales are keeled. The head is small and similar in thickness to the neck. The ***labials (lips) are yellowish or greenish with bold black vertical bars***. Both species of ribbonsnakes have clean white labials and are more slender than a gartersnake. The

Found in prairies and other grassy areas near water.

Close-up showing the thick black labial bars. Common Gartersnake has thin or none here.

This species was formerly divided into two subspecies.

Common Gartersnake has no black or just thin markings on its labials.

Juveniles have the same colors and pattern as adults.

Diet: Frogs and tadpoles, salamanders and their larvae, toads, earthworms, fish and beetles.

Courtship, Mating & Eggs: Plains Gartersnakes mate during April and May. A male will align his body beside the female, and when she lifts her tail, he will insert his hemipenis. After ten or fifteen minutes of copulation, they are done. Often there is more than one male, but male number two might be out of luck. The previous male may have left a mucous plug in her cloaca that renders her unable to mate for several days. In August or September the female will give birth to 5 to 97 young.

Range Notes: Ranges south to include northern Illinois and Iowa. Also Alberta, Saskatchewan, Manitoba, the northwest half of Montana, the Dakotas and south into eastern Colorado and northwest Oklahoma.

Red-sided Gartersnake (subspecies of Common Gartersnake) *Thamnophis sirtalis parietalis*

Prairie, but usually in the vicinity of water. Also, fields, grasslands and farmlands.

Nature Notes:

A Red-sided Gartersnake's first response to danger would be a quick escape into cover. That failing, it would coil and strike at an enemy. If grabbed, it would bite and release a foul smelling musk. Only annoying, but sometimes it works for them.

snakes

This snake is a subspecies replacing the Eastern Gartersnake in western Minnesota.

Description: Length is 18 to 26 inches. Species record is 54 inches.

The Red-sided Gartersnake is one of the subspecies of the Common Gartersnake found in our area. As its name states, this snake has ***wide red bars or spots on its sides between the dorsal stripe and the side stripe that are easily visible from a distance*** against the black or dark brown ground color. The scales are keeled. The labial (lip) scales are nearly unmarked and are light yellow, cream colored or with a slight greenish tint. The top of the head is dark with a pair of tiny yellow parietal spots.

Newborn snakes are similar in color and pattern to adults.

Diet: Mainly earthworms and amphibians, but baby ground nesting birds or their eggs,

Lots of red shows between the scales and the black spots on the Red-sided Gartersnake.

This snake has less red and is an intergrade with Eastern Gartersnakes.

fish, small mammals and invertebrates are also on the menu.

Courtship, Mating & Eggs: Mass numbers of Red-sided Gartersnakes sometimes share a hibernaculum and when the snakes emerge, they are ready to breed. Competition among males to breed with a receptive female is fierce, so some dirty tricks evolved. Some males will secrete an odor mimicking a female to distract other males. Sometimes a male, after successfully mating, will leave a mucus-like plug in the female's cloaca so no other male can breed with her for a few days.

In August or early September, the female will have 3 to 80 (most likely 10 to 40) young.

Range Notes: Western Minnesota, most of Iowa and west to Nebraska, Kansas and Oklahoma. Northwest to parts of the Dakotas, Montana and Manitoba; west to British Columbia.

snakes

Eastern Gartersnake (subspecies of Common Gartersnake) *Thamnophis sirtalis sirtalis*

Habitat includes water or in the vicinity of water, including forests, prairies, meadows, marshes or vacant lots.

Nature Notes:

Gartersnakes are quick to escape into tall grass or brush if danger threatens. Sometimes they coil and spread out the body to look bigger. Like many other snakes, when grabbed, they bite and release a foul smelling musk from glands at the base of the tail. Neither defense is more than mildly annoying, but it often works.

A very adaptable snake that can thrive in forests, forest edges, meadows, marshes, bogs and vacant lots.

Description: Length is 18 to 26 inches. Record is 54 inches.

The Eastern Gartersnake is the main subspecies in our three states. The scales are keeled. Three yellow stripes are usually prominent against a background of black, brown, or olive. The *side stripes are located on scale rows 2 and 3. Many appear spotted or checkered, having two rows of alternating black spots between the dorsal and lateral stripes.* The top of the head is dark and there is *little or no marking on the labial (lip) scales.*

Juveniles are miniature versions of the adults.

Diet: Eastern Gartersnakes feed on mostly cold-blooded prey, such as frogs, toads, salamanders, fish and earthworms.

A Green Frog becomes lunch for this gartersnake (right).

This gartersnake probably just made a meal of a frog or toad.

The snake's tongue is used to help sense the environment.

Photographed in the intergrade zone, this snake shows some traits of both Red-sided and Eastern.

Courtship, Mating & Eggs: Eastern Gartersnakes are ready to breed when they emerge from hibernation. The result is intense competition among males and a couple dirty tricks. Some males secrete an odor mimicking a female to distract other males. Sometimes a male, having successfully copulated with a female, will leave a plug of mucus-like substance inside her cloaca that insures no other males can breed with her for several days.

In August or early September, the female has anywhere from 3 to 80 (most likely 10 to 40) young.

Range Notes: The Eastern Gartersnake is found throughout our three states, except in western and southwestern Minnesota where it is replaced by the Red-sided Gartersnake. Intergrades between the two subspecies exist in a broad diagonal from northwest to southeast Minnesota. Intergrades of the Chicago Gartersnake, *Thamnophis sirtalis semifasciatus* and Eastern Gartersnake exist in extreme southeastern Wisconsin.

snakes

Lined Snake

Tropidoclonion lineatum

Prairies, fields, grasslands, pastures, vacant lots, parks and cemeteries.

Nature Notes:

Lined Snakes are secretive, hiding under flat rocks and logs during the day, and prowling for food at night. A heavy rain will also bring them out.

They spend winter below the frost line in deep cracks in rocks or in burrows.

A docile snake that would only rarely bite a person.

When threatened, the Lined Snake will hide its head under its body coils and release its foul smelling musk. It might flatten its body and raise its curled tail to distract a predator.

A small stout-bodied prairie snake with three stripes similar to a Gartersnake but with a diagnostic double row of half-moon black spots down the belly.

Description: Length is 9 to 15 inches. Record is 21 ½ inches.

The Lined Snake is ***brown, olive or gray colored with a yellowish stripe down its back and one more stripe on each side.*** The middle stripe may be flanked on each side by a row of black spots, and another row of black spots may be above the side stripe. The scales are keeled. Compared to gartersnakes, the head and eyes are smaller, and its ***belly is pale with two rows of black half-moon spots*** running down the middle.

The juveniles are, surprisingly enough, even less colorful than the adults, except for the distinct half-moon belly spots.

This Lined Snake is coiled defensively with a flattened body and a raised curled tail.

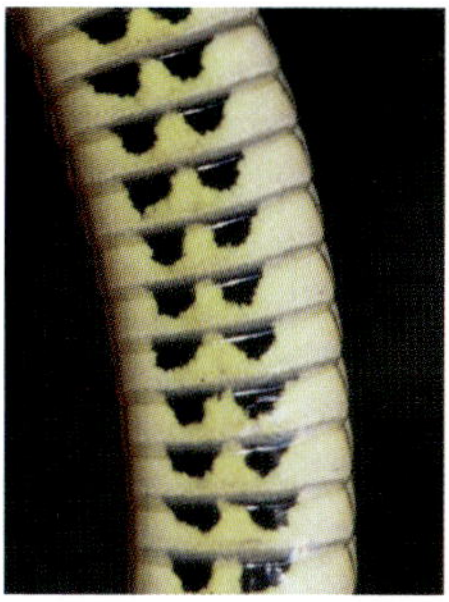

Belly with its double row of black half-moon spots.

Courtship, Mating & Eggs: Lined Snakes mate in the fall, unlike most other snakes, and the female retains the sperm until spring when her eggs are fertilized. In August or early Septmber, she will bear 2 to 17 young. The males mature in less than a year and the females are ready to mate in their second year.

Diet: Mostly earthworms but also soft-bodied insects and insect larvae. Earthworms are seized and swallowing begins immediately so the worm has no chance to retreat into its burrow. The worm is doubled up going down the snake's throat.

Range Notes: Two isolated areas in our region: Extreme southwestern Minnesota and three counties in south central Wisconsin. Scattered populations south to Texas and New Mexico.

Timber Rattlesnake

Crotalus horridus

River bluffs with southwest or south facing rocky outcrops and their surrounding hillsides and valleys.

Nature Notes:

Timber Rattlesnakes start emerging from hibernation in April, and their numbers peak in May. In June the unmated females and males are dispersing down into summer habitat, while the gravid (pregnant) females remain up on the bluff.

Timber Rattlesnakes have a pair of long, hollow fangs that fold back along the roof of their mouth. They are ambush predators that sit by a fallen log or at the base of a tree where rodents frequently travel. Even in complete darkness, the heat-detecting pits reveal the target for an accurate strike. The forked tongue and Jacobson's organ are then used to track down the meal.

A large pit viper, this snake is an ambush predator which strikes its prey and then tracks it down.

Description: Length is 36 to 60 inches. Record is 74½ inches.

The Timber Rattlesnake is venomous and has *a wide head, distinct from the neck. Its eyes have elliptical pupils which, in the sunlight, appear as vertical slits.* Between each eye and nostril is a heat-sensing pit. The *top of the head is unmarked.* The scales are keeled. The *heavy body has narrow, jagged black or dark brown bands* over a ground color of yellow, tan, brown, rusty brown or occasionally gray. *The tail is black and tipped with a tan rattle.*

Newborn Timber Rattlesnakes are gray with dark bands until, a few days later, they shed for the first time. Then they have the same colors and pattern as adults.

South-facing rock outcrops are the place to look for Timber Rattlers in our region.

Notice the facial pit and vertical pupil; traits of rattlesnakes.

Habitat is the river bluffs of southeast Minnesota and southwest Wisconsin.

Timber Rattlesnake showing its rattle and black tail.

Diet: Chipmunks, squirrels, ground squirrels, small rabbits, mice and voles.

Courtship, Mating & Eggs: Courtship may be preceded by combat between two males, as they raise up high to push each other over for dominance. Courtship and mating are usually in late summer and through September. The female retains the sperm until spring when her eggs are fertilized. In late August or September, she bears 3 to 18 young and they will stay together for 10 to 14 days. The female won't have young again for three or four years.

Juvenile Timber Rattler showing its gray color.

snakes

Range Notes: From our region range extends to northeast and southeast Iowa, western, southern, northcentral and eastern Illinois; southwest to east Texas, east to northern Florida and north to New Hampshire.

Eastern Massasauga

Sistrurus catenatus

River bottom forests, swamps, and wet prairies. May move to nearby fields in the summer.

Nature Notes:

Heat-sensing pits help detect and accurately strike warm-blooded prey in complete darkness, and the hollow fangs inject venom. Its tongue and Jacobson's organ are employed to track its meal.

Massasaugas hibernate individually in crayfish burrows or other burrows, and they emerge with spring flooding.

Not an aggressive snake and uses its protective camouflage to avoid being bothered. When threatened, it will coil and vibrate its tail, although its small rattle is not loud, resembling the buzz of an insect.

A Wisconsin DNR researcher was accidentally bitten on the thumb by a Massasauga. Lucky for her, it was a dry bite. The snake decides how much venom, if any, to inject.

A pit viper that doesn't use a den for hibernation but utilizes crayfish or mammal burrows.

Description: Length is 20 to 32 inches. Record is 39 ½ inches.

The Eastern Massasauga is a venomous rattlesnake with a ***heavy body and a small rattle.*** The head is large, widened at the rear and clearly larger than the neck. There is a pit between the eye and nostril, and the ***eye has a vertical pupil.*** The scales are keeled. The body is ***gray, brown, or dark brown with darker blotches.*** Sometimes the blotches are edged with white. ***The blotches turn to bands on the tail.***

Juveniles are similarly colored, but the rattle is only a single prebutton on newborn snakes.

Diet: Voles and deer mice primarily but also frogs, snakes, birds and bird eggs.

Courtship, Mating & Eggs: Massasaugas breed in fall and spring, although many

Note the Massasauga's vertical pupil and rattle; both are distinguishing traits of rattlesnakes.

Juvenile with a prebutton rattle.

females only breed every other year. The 5 to 20 young are born in August or early September and are capable of using their fangs and venom.

Range Notes: The Massasauga is found in most of Lower Michigan. It is uncommon in the central sands area of Wisconsin and west and south. It is not in Minnesota until one swims across the Mississippi. From our area, most of southeast Ontario with isolated populations in New York and Pennsylvania; west to Iowa and Missouri.

snakes

Glossary

Aberrant: Coloration different than normal for the species

Adhesive disc: Enlarged toe tip of a frog in the treefrog family.

Amnion: A fluid-filled membrane surrounding the embryo of a reptile, bird, or mammal.

Amplexus: The sexual embrace of male and female frog or toad.

Anal plate: The enlarged scale (single or divided) in front of or covering a snake's vent.

Annuli: Concentric annual growth rings evident on the scutes of some turtles.

Anterior: Towards the front.

Anuran: A frog or toad.

AOR: "Alive on road" when recording data on an amphibian or reptile. ?

Arboreal: Living in trees.

Barbel: A small fleshy skin projection on the chin or neck of a turtle.

Boss: A large bump between the eyes of some toads.

Bridge: The part of a turtle's shell connecting the carapace and plastron.

Brille: The clear scale covering the eye of a snake.

Carapace: The upper part of a turtle's shell.

Caruncle: Temporary, sharp tooth-like structure on the snout of a hatchling turtle.

Cirri: A pair of tiny points projecting downward from upper lip of some salamanders.

Cloaca: The internal chamber that holds wastes or reproductive materials.

Costal grooves: Vertical grooves on the sides of salamanders.

Crepuscular: Most active at dawn or dusk.

Crests: Ridge-like structures on top of the head of some toads.

Crustacean: Aquatic arthropods with hard shells, such as a crayfish.

Cryptic: Camouflage coloration.

Cusp: A pointed tooth-like projection.

Disjunct: A population geographically separated from the main range of a species.

Diurnal: Active during daylight.

DOR: "Dead on road" when recording data on an amphibian or reptile.

Dorsal: Referring to the back.

Dorsolateral ridge: A raised ridge along the upper sides of some frogs.

Dorsum: The upper surface.

Ecdysis: The shedding of the skin.

Ectotherm: An animal that gains its body heat from the environment.

Eft: Terrestrial stage of the newt.

Egg tooth: A temporary tooth used by snakes and lizards to slice the egg shell.

Glossary

Extirpated: All individuals of a species are no longer found in a specific region.

Facial pit: The heat-sensitive organ located between eye and nostril of pit vipers.

Hemipenes: The paired copulatory organs of male lizards and snakes.

Herbivorous: Eating plants, not animals.

Herp: An amphibian or reptile.

Herpetofauna: The amphibians and reptiles of a specific area.

Herpetology: The study of amphibians and reptiles.

Hibernaculum: The den or place an animal spends the winter.

Intergrade: The offspring of two different subspecies.

Keratin: A tough protein composing the outer covering of reptile skin and scales.

Labial: Referring to the lip area.

Medial: Refers to the middle or midline.

Middorsal: Referring to the middle of the back.

Morphology: The appearance, form, or structure of an animal.

Nasolabial groove: A tiny groove extending from nostril to mouth of some salamanders.

Neotenic: Describes a salamander that retains larval features as an adult.

Ocelli: Dark or light edged circular spots.

Oviparous: Egg-laying.

Ovoviviparous: Young develop in the body, but nurtured by yolk sac not placenta.

Parotoid glands: A pair of large poison glands behind a toad's eyes.

Plastron: The bottom part of a turtle's shell.

Plates: Large reptile scales, especially on a turtle shell, but also on the belly of snake.

Posterior: Toward the rear of an animal.

Prebutton: The keratin structure at the end of a new-born rattlesnake's tail.

Release call: A call given by a male anuran when clasped by another male.

Relict population: A remaining group of a species, left isolated by a shrinking range.

Reticulate: A netlike pattern of linear markings.

Rostral: Scale on the tip of the snout.

Scale: An outgrowth of keratin from reptile skin

Scute: A large scale, such as on a turtle shell or a snake's belly.

Sexual dimorphism: Difference in form between the sexes.

Snout-vent length: Length of an amphibian or reptile not including the tail.

Spermatophore: A gelatinous mass capped by sperm deposited by a male salamander to be picked up by the female.

Glossary

Spiracle: A short tube that discharges water from a tadpole's gill chamber.

Subspecies: A subdivision of a species when two or more groups of the species differ slightly in color or morphology.

Terrestrial: living on land.

Tetraploid: An animal with double the normal number of chromosomes.

Tubercle: A small rounded skin projection.

Tympanum: External eardrum of most anurans and turtles.

Vent: The external opening of the cloaca.

Ventral: Refers to the belly.

Ventrals or ventral plates: The wide belly scales of snakes.

Vertebral: The middle of the back.

Viviparous: Bearing live young that were nourished through a placenta.

Internet Resources

Adorama *www.adorama.com*—camera equipment.

Amphibians & Reptiles of Iowa, Minnesota & North Dakota *www.herpnet.net*—Herpetology and species in these three states.

B&H Photo *www.bhphotovideo.com*—camera equipment.

Center for North American Herpetology *www.cnah.org*—herpetology information.

Field Herp Forum *www.fieldherpforum.com*—field herping discussions.

Global Amphibian Assessment *www.globalamphibians.org*—status of amphibians.

Herpmapper *www.herpmapper.org*—submit sightings and keep a lifelist.

Herpetology of Minnesota *www.herpnet.net*—amphibians and reptiles of Minnesota.

iNaturalist *www.inaturalist.org*—submit sightings, crowd-sourced identification.

League of Conservation Voters *www.lcv.org*—environmental voting records.

Midwest Tongs *www.tongs.com*—snake hooks and tongs.

Minnesota Herpetological Society *www.mnherpsoc.org*—herpetology.

NatureSound Studio *www.naturesound.com*—frog and toad calls.

Orianne Society *www.oriannesociety.org*—conservation of herps & habitats.

Society for the Study of Amphibians and Reptiles *www.ssarherps.org*—herpetology.

Facebook Group — *Amphibians & Reptiles of Minnesota*

Facebook Group — *Michigan Reptile & Amphibian Identification*

Photo Credits

T=top	R=right
M=middle	L= left
B=bottom	I=inset

All the amazing photos in this guide were taken by the author, **Allen Blake Sheldon**, except those listed below...

Anne Elliot: 77 T, BL

James Gerholdt: 67 RM, LM

Jim Harding: 178

Jeff LeClere: 41 T, 66, 67 T

Peter Paplanus [2ndPeter at www.Flickr.com]**:** 43 B, 51 B, 56, 57, 85 BL, 139 T, 140, 141 B, 155 T, 189 T

Todd Pierson [twpierson at www.Flickr.com]**:** 179

Public Domain/Creative Commons: 49 T (HARGLE, WIKIMEDIA COMMONS), 49 B (ENTOBILL, WIKIMEDIA COMMONS), 67 B (CEASOL, WIKIMEDIA COMMONS), 141 T (DON BECKER, WIKIMEDIA COMMONS)

Sparky Stensaas [www.ThePhotoNaturalist.com]**:** 60 I, 72 T, 73 B, 75 BL, MR, 76 B, 81 B, 89 BL, 91 TR, 97 TL, 106 B, 108 MAIN, B, 109 MR, 124 INSET R, 135 M, B, 137 BL, 149 B, 187 BR

WisconsinTrailGuide.com: 109 TL

Titles of Interest

Christoffel, R., R. Hay, and M. Monroe. 2002. *Turtles and Lizards of Wisconsi*n. Madison, WI: Wisconsin Department of Natural Resources.

Christoffel, R., R. Hay, and L. Ramirez. 2000. *Snakes of Wisconsin*. Madison, WI: Wisconsin Department of Natural Resources.

Chistoffel, R., R. Hay, and M. Wolfgram 2001. *Amphibians of Wisconsin*. Madison, WI: Wisconsin Department of Natural Resources.

Conant, Roger, and Josepht T. Collins. 1998. *Peterson Field Guides: Reptiles and Amphibians: Eastern/Central North America*. New York, NY: Houghton-Mifflin.

Gibbons, Whit. 2017. *Snakes of the Eastern United States*. Athens, GA: The University of Georgia Press.

Harding, James H. 1997. *Amphibians and Reptiles of the Great Lakes Region*. Ann Arbor, MI: University of Michigan Press.

Harding, James H. and J. Alan Holman. 1992. *Michigan Frogs, Toads and Salamanders*. East Lansing, MI: Michigan State University Museum.

Harding, James H. and J. Alan Holman. 1990. *Michigan Turtles and Lizards*. East Lansing, MI: Michigan State University Museaum.

Holman, J. Alan, James Harding, M. M. Hensley and G. R. Dudderar. 1989. *Michigan Snakes*. East Lansing, MI: Michigan State University.

Johnson, Tom. 1987. *The Amphibians and Reptiles of Missouri*. Jefferson City, MO: Conservation Commission of the State of Missouri.

LeClere, Jeffrey B. 2013. *A Field Guide to the Amphibians and Reptiles of Iowa*. Rodeo, NM: ECO Herpetological Publication & Distribution.

Moriarty, John J. and Carol D. Hall. 2014. *Amphibians and Reptiles in Minnesota*. Minneapolis, MN: University of Minnesota Press.

Moriarty, John J. 2004. *Turtles and Turtle Watching for the North Central States*. St. Paul, MN: Minnesota Department of Natural Resources.

Powell, Robert; Conant, R. and J. T. Collins. 2016. *A Field Guide to Reptiles and Amphibians of Eastern and Central North America*. New York, NY: Houghton Mifflin.

Shaw, John. 1987. *John Shaw's Closeups in Nature*. New York, NY: Amphoto.

Sheldon, Allen Blake. 2006. *Amphibians & Reptiles of the North Woods*. Duluth, MN: Kollath+Stensaas Publishing.

Trauth, S. E., H. W. Robison, and M. V. Plummer. 2004. *The Amphibians and Reptiles of Arkansas*. Fayetteville, AR: University of Arkansas Press.

Vogt, R. C. 1981. *Natural History of Amphibians and Reptiles of Wisconsin*. Milwaukee, WI: Milwaukee Public Museum.

Additional References

Blaustein, A. R. 1994. *Amphibians in a Bad Light.* Natural History. October: 32-39.

Parmelee, J. R., M. G. Knutson, and J. E. Lyon. 2002. *A Field Guide to Amphibian Larvae and Eggs of Minnesota, Wisconsin and Iowa.* Washington D.C.: U. S. Geological Survey.

Relyea, R. A. 2005. *The Lethal Impact of Roundup on Aquatic and Terrestrial Amphibians.* Ecological Applications 15 (4): 1118-1124.

Sanders, R. 2006. *Pesticide Combinations Imperil Frogs.* Lawrence, KS: Center for North American Herpetology News Release, 10/17/06.

Sheldon, A. B. 1995. *Photographing Wild Reptiles.* Reptile & Amphibian Magazine. July: 76-83.

Sheldon, A. B. 1997. *Night Photography of Frogs & Toads.* Reptile & Amphibian Magazine. March: 70-73.

Sheldon, A. B. 2004. *Wading for a Photo Op: Secrets of Shooting Frogs at Night.* Big River Magazine. March-April: 18-19, 25-27.

Sheldon, A. B. 2005. *Northern Watersnake.* Big River Magazine. Sept.: 28-31.

Sheldon, A. B. 2006 *Meet the Mudpuppy.* Big River Magazine. May-June: 22-25.

Sheldon, A. B. 2006. *Shooting the Nightlife.* Reptiles Magazine. December: 36-43.

Sheldon, A. B. 2007 *Rattlesnakes along the Upper Mississippi.* Big River Magazine. May-June: 26-29.

Sheldon, A. B. 2014. *Turtle-spotting on the Mississippi.* Big River Magazine. May-June: 32-36.

Frog and Toad Calls

Elliot, L. 1992. *The Calls of Frogs and Toads: Eastern and Central North America.* Post Mills, VT: Chelsea Green Publishing.

Kellogg, P. P., A. A. Allen, and T. Wiewandt, 1982 *Voices of the Night: The Calls of the Frogs and Toads of Eastern North America.*

Index

Quick Herp Finder